RAGS make paper,
PAPER makes money,
MONEY makes banks,
BANKS make loans,
LOANS make beggars,
BEGGARS make
                    RAGS.

—Anonymous, c. 18th century

PAPER-MAKING

BY JULES HELLER

WATSON-GUPTILL PUBLICATIONS/NEW YORK

To my grandchildren:
Sonia Louise Davis and
Evan Alexander Davis
—with love—and, in friendship

Jules Heller received his B.A. from Arizona State University in Tempe, his M.A. from Columbia University in New York City, and his Ph.D. from the University of Southern California. He was for many years the Dean of the College of Fine Arts (as well as Professor of Art) at Arizona State University, he has in the past served as Founding Dean of the Faculty of Fine Arts at York University in Toronto, and Founding Dean of the College of Arts & Architecture at The Pennsylvania State University. He has traveled throughout the world to places such as Sri Lanka and Thailand as visiting professor.

The artist/writer has written a book, *Printmaking Today*, in addition to many articles in various magazines, including *Artist's Proof* and *Impression,* of which he was Executive Editor from 1957 to 1958. He is listed in *Who's Who in America, Who's Who in American Art,* and *Who's Who in the World.*

Mr. Heller has participated in solo and group exhibitions including ones at the Gallery Pascal in Toronto, the Arizona State Capitol Building, the Martha Jackson Gallery in New York City, and the American Color Print Society in Philadelphia. His work is in numerous permanent collections, both private and public—from the Toronto Dominion Centre to the University of New Mexico.

Jacket design by Bob Fillie
Front Jacket: Paper casting by Jennifer Place.
Photo by Donald Holden.
Back Jacket: Paper Relief - Kozangrodek by Frank Stella, 1975.
Courtesy Tyler Graphics, Bedford, New York.

First published in paperback 1997 in New York
by Watson-Guptill Publications,
a division of Billboard Publication, Inc.,
1515 Broadway, New York, N.Y. 10036

Library of Congress Cataloging in Publication Data
Heller, Jules.
    Papermaking—the white art. Pbk. Ed.
    Bibliography: p.
    Includes index.
    1. Papermaking and trade. I. Title
TS1105.H54   676'.22   78-691
ISBN 0-8230-3842-4

Manufactured in U.S.A.

1 2 3 4 5 6 7 8 9 10/06 05 04 03 02 01 00 99 98 97

ACKNOWLEDGMENTS

Sources and credits for the photographs are acknowledged elsewhere; my special thanks to all of the photographers for allowing me to utilize their skills, to the artists for their unbelievable cooperation—present and past—to dealers, collectors, galleries, and museums for courtesies extended, and to hosts of friends and acquaintances for offering more materials than could ever be accommodated in one volume. I wish it were otherwise.

For particular kindnesses rendered, I wish to thank (in alphabetical order) Eishirō Abe, Living National Treasure of Japan; Professor Suzanne Anker of Washington University; John Babcock, artist; Laurence Barker, artist-papermaker of Barcelona, Spain; Mary Welsh Baskett, Curator, Museum of Contemporary Crafts, New York; Dr. Neil Berman of Arizona State University; Don Farnswoth of Farnsworth & Co.; Nancy Genn, artist; Professor Walter Hamady of the University of Wisconsin in Madison; Harold H. Heller, senior research consultant to Appleton Mills; Charles Hilger, artist; Douglass Morse Howell, our unofficial Living National Treasure of papermaking, for his inspirational letters; Professor Bill Jay of Arizona State University; John Koller, proprietor of the paper mill, HMP; Elaine Koretsky, proprietor of Carriage House Handmade Paper Works; Roberta Loach, editor/publisher of *Visual Dialog*; Professor Winifred Lutz of Yale University; Henry Morris, proprietor of the Bird & Bull Press; John Mason, proprietor of the Twelve by Eight Press in Leicester, England; Harold Persico Paris, sculptor-papermaker; John R. Peckham, Research Fellow, Institute of Paper Chemistry; J. Norman Poyser, of the Pulp and Paper Research Institute of Canada; Dr. Peter Serjeant of West Virginia Pulp & Paper Co.; Asao Shimura, papermaker of Tokyo; Ann Tullis, Director of the Institute for Experimental Printmaking; Joe Wilfer, proprietor of Upper U.S. Paper Mill; J. Tuzo Wilson, Director General, Ontario Science Centre, Toronto; James Yarnell, proprietor of Oak Park Press and Paper Mill; and Professor Joseph Zirker of San Jose City college.

To Jennifer Place, Donald Holden, and Ellen Zeifer of Watson-Guptill, my gratitude for assistance beyond what authors normally receive.

My special thanks to Elsie Sawyer for typing this manuscript and to Gloria Heller for her remarkable insights, support, and power of perception.

CONTENTS

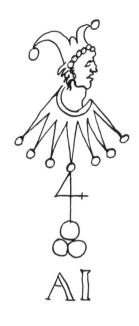

Part II. Theory and Some Practice 91

PREFACE

*Let us begin by enumerating those qualities expressing
the aptitude or inaptitude of a thing to be affected
in a certain way. They are as follows: to be apt or inapt
to solidify, melt, be softened by heat, be softened
by water, be comminuted, impressed, moulded, squeezed,
to be tractile, malleable, fissile, be cut, be viscous
or friable, compressible or incompressible, combustible
or incombustible, apt or inapt to give off fumes.*

—Ascribed to Aristotle

There have been days, in the course of working on this manuscript, when I believed it was absolutely necessary that everything come to a screaming halt to allow me to enroll in courses and take further degrees in art, botany, polymer chemistry, geography, history, materials science, forestry and wildlife, wood technology, physical chemistry, general chemistry, elementary woodwork, general agriculture, fiber science, chemical engineering, and fluid mechanics.

There were other days when I convinced myself that if man made paper for approximately 2,000 years without benefit of machines and present-day industrial research and developmental organizations, and some of that paper still exists, why should I fill my head with facts no one ever knew until quite recently?

I took two aspirins and quietly went to bed to mull it all over again. . . .

The beginning is the end and vice versa, especially as concerns the making of handmade paper; it has been a reversible process for millennia. It is only after one has completed such a work that reality percolates to the surface: you are shocked that you consulted hundreds of individuals, few of whom agreed with each other, all of whom were absolutely firm in their beliefs that they and they alone knew "the way"; and, I confess, I admire all of them for their steadfastness of purpose, their oneness of vision, their unshakable conviction.

It has been my privilege to have visited, at some time, almost all of the professional mills in the United States and Canada and to have spent a brief time with each of the proprietors-artists-papermakers.

Despite the seemingly insurmountable problems, real and imagined, that we individually and collectively face, we are witness (and many are participants) to the revolution in paper. This revolution will no longer allow a sensitive individual to examine a figure without

equally inspecting the ground on which it sits or in which it is partially or completely submerged, or of the substance itself—paper.

You will find contradictory and sometimes controversial areas of information contained in this book. Please accept this as the norm in a rapidly advancing, handmade industry. Even that last phrase reads in a most peculiar fashion. How can you have a handmade industry? Yet, strangely enough, we do.

And, in the stage of development in which we now find ourselves, in our desire to be purists who use no additives of any sort—no dyes, no buffers, nothing but the rags from our backs and those of our friends, plus that which is provided us by nature and is free and wild; or in our desire to become more scientific than the paper chemists and the super-specialists who insist on careful records of pH, exact and repeatable measurements of any and all additives, careful timing of beating, etc., etc., to provide us with the pH neutral sheets we demand; or in some place on the imaginary spectrum I have just drawn, we can and all will be accommodated in the process of handmade papermaking.

In crude form, we appear to be divided into two groups: those who see in the process of handmade papermaking an end-product for printing, printmaking, drawing, calligraphy, and watercolor painting, and those who view the new paper revolution as a medium of artistic expression in its own right, without reference to other mediums. Perhaps in some ways, that crude division is unfair, because there are papermakers who are artists and there are artists who are more technically oriented than technicians. Perhaps we are looking at two sides of the same coin. Theory and practice. Thesis and antithesis. The right half and the left half of the brain.

My goal in writing this book has been to attempt to bring before three separate audiences sufficient

information, in word and image, to enable all of you to make handmade paper: the child in all of us (of any age) who will want to try this new-old process once, twice, or more times, using the tools and materials found in your home or apartment; the student of any age who is seriously interested in the making of handmade paper and who has access to or will make or purchase materials and equipment to further that interest; and, finally, the individual who—"hooked" on handmade paper—will take that quantum leap, financially and otherwise, and purchase professional equipment, using it to satisfy whatever deep, private, inner urges need satisfying.

It has, further, been my intention to provide something "old," something unexpected in each chapter of this book—in the hope that the reader, whenever he or she opens the book, will discover something useful. Hopefully, that tiny "nugget" may stimulate a pride of visual ideas, unlock a flood of projects, encourage the unfolding of a hitherto secret creative growth and development.

Additionally, I wanted to try and reconcile the seemingly conflicting aims between those whose sole desire is to make perfect sheets of paper as carriers for the images of artists and printers, and the myriad individuals, all over the world, who see in paper a medium in its own right, a tool with which to make cast paper works, or other three-dimensional works of paper, by using pure paper pulp in new and meaningful ways.

I believed, and I still do, that one group cannot exist without the other; that both, whether they are aware of it or not, need to cooperate fully and openly, if the revolution in papermaking is to be fully realized. Further, to encourage growth and forward progress, it is essential that both groups become familiar with the history of papermaking through organizations devoted to such study. For example, is it well known, in the western

world, that cast paper works, (presently regarded as new and avant-garde—including bas-reliefs and all sorts of decorative ceiling treatments) were quite popular in eighteenth-century England and Ireland? That many experiments in new materials for making sheets of paper have been going on for generation upon generation?

"If I choose to style myself a 'gentleman papermaker,' please do not take it that I give the term undue emphasis and that I wish to imply that not all papermakers are gentlemen. However, there are some devils among them as there are among printers." (Harrison Elliott, N.Y.P.L., p. 111.)

In the course of working on the manuscript for this book, I talked or corresponded with amateur and professional papermakers, paper dealers, chemical engineers, and paper scientists—in such rare atmospheres as Departments of Delignification—senior research consultants for paper mills, artists who worked with handmade paper, professors who offered courses in the field of handmade paper, limited edition printers and publishers of art magazines, and the clutch of persons in North America who support themselves entirely by the sale of handmade and custom papers to others. On rereading this passage, I may have left out numbers of people. If so, my apologies; perhaps I have arrived at one of those stages in life where forgetfulness is rampant.

Although I feel equivocal about storytellers, I cannot allow the following anecdote to go unrecorded. In the course of one of my many recent conversations and interviews, I asked a New York paper enthusiast and dealer, Steven Steinberg, for his favorite real-life story related to handmade papermaking. This was his response:

"Well, this is almost embarrassing. There was some publicity about an Irish paper mill that I discovered, and the interviewer asked me questions about all the various papers I had found all over the world. I related the incident about this little paper mill in Ireland, showed some samples, and described the paper: a handmade sheet, light blue in character, with red and blue fibers running through it. So, the article was published and I was very excited. This was two years ago. I told everybody about the paper, trying to promote its use, and two very young, attractive women came into the shop. I was taken aback by them, and we sat and talked a little bit and they kept nudging me about this blue paper with the red and blue fibers. I kept showing them other papers, you know, and they would love them, but they just kept coming back to the blue paper (that I was still to receive from Ireland). Finally, the two women looked at each other, and they looked at me. They reached into their wallets and respectively took out two badges, saying United States Secret Service, and I almost fainted!

"I thought, 'My God. I paid my income tax.'

"She said, 'We have to be very candid; we are not really interested in any of the other papers. We are solely interested in that blue paper.'

"The description that I gave them sounded exactly like the kind of paper suitable for counterfeiting. They told me that I was not allowed to own paper with those red and blue fibers running through. There seems to be some sort of statute.

"I tell you, I needed a strong drink after that."

In conclusion, despite the assistance of a multitude of people—some of whom are named in the body of the work—I take full responsibility for any errors of omission or commission found in this book, and sincerely hope, in the course of time, that I will have the privilege of revising or even rewriting my thoughts on the subject as errors are discovered and new processes and artists discover handmade paper.

I cannot resist a few words of advice: beware the seemingly simple and easy approaches. All are arduous, filled with hidden pockets of resistance in the face of perversity, time-consuming, back-breaking, and harmful to a normal social life.

A micro-list of what would appear to be facile activities includes the golf swing of the professional "on the tour," the fly caster's line describing a lazy figure-8 roll forward and backward in space, and a Zen poem, along with making handsome paper.

J.H.
Scottsdale, Arizona
September, 1977

*A centuries-old view
of the papermaking process.
Anonymous woodcut.*

INTRODUCTION. WHY MAKE YOUR OWN PAPER?

Various the papers, various wants produce,
The wants of fashion, elegance, and use,
Men are as various: and if right I scan,
Each sort of paper represents some man.

—Attributed to Ben Franklin, 1787

With the possible exception of one of Dard Hunter's contributions to papermaking, it is difficult to find, in one volume, the complete story of the so-called "white art" for artist-papermakers from today's vantage point—unless the reader has access to a major library collection of rare and limited edition books on the subject.

Curious. Throughout the centuries, to this very day, people have taken paper for granted. It is regarded as one of the givens of society, as ubiquitous as rain, smog, motherhood, or oleomargarine. Being so obvious, it has long been invisible. If requested to "think paper," most individuals will meditate on a sheet of white paper. Further, it is widely believed that pure, white paper (as with a certain brand of well-advertised soap) is the omega of papermaking.

How do you define the color, white? What images, what associations come to mind? The albuminous material surrounding the yolk of an egg; the fifth circle of an archery target; the purity and cleanliness of a well-scrubbed, white-enameled kitchen sink; the virgin-whiteness of a wedding gown; great masses of flour, sugar, and snow; Snow White and her seven little men; the white part of the eyeball; hooking a good-sized white bass; the silvery white of the birch; whitecaps on duck-egg blue water; whitewash (political and the Mark Twain variety); white elephants, both literal and figurative; the white-face of mimes and clowns; whitefish (smoked) for Sunday brunch; a White Friar and Whitefriars in Fleet Street, London; the American bald eagle; the white heat of anger and the fear-provoking White Horde; white-hot metal and the 374 foot White Horse of Saxon fame; a certain eighteenth century colonial mansion in Washington, D.C.; Kipling's unfortunate "white man's burden"; white nebula and the white noise of electronic music; a Canadian winter white-out; the White Rose of York and White Russians; January white

sales and a leaping white (silver) salmon on the Kaniapiskau River; white sauce for madame and a man-eating shark for monsieur; white slavery and white supremacists seen against the background of the White Terror of eighteenth century France; white tie and tails along the Great White Way; Melville's whale and whitewings (streetsweepers)—to list a double-clutch of words found in the nearest dictionary. But enough. Let us leave this intriguing digression with the disturbing thought that *white,* in the eastern world, carries with it vast numbers of associations quite *other* than western man's conceptions.

WHY NOT BUY MACHINE-MADE PAPER

Many users of expensive papers for drawing, printmaking, and watercolor painting have been deluding themselves for years in believing they have been buying and using handmade papers when, in fact, they have been purchasing *mould-made* papers (made by machine) containing varying percentages of rag content. (The differences between machine-made, mould-made, and handmade papers will be explained later.)

Show a prospective customer a deckle edge and whisper, "rag content," and you have a buyer—albeit an ignorant one. Yet, there is nothing wrong with mould-made paper; some of my best prints were and are made on it, and many of my best friends—to overwork an overworked cliché—still use Antique Laid, German Copperplate, English Etching, Arches, Rives BFK, and other mould-made paper for prints, drawings, and watercolors and are delighted with their results. Compare any mould-made paper with handmade, however, and a different story results, which will unfold as we continue.

In the not-too-distant past, there was a plentiful supply of inexpensive handmade paper. Mills employed highly skilled craftsmen to replenish their stocks of paper for

sale around the world. Unfortunately, the overwhelming majority of these old paper mills have closed down, victims of the technological revolution, for lack of young recruits to the craft, rising costs, inflation, and steadily decreasing profit margins.

In 1886, Charles Thomas Davis wrote, "There is now so little handmade paper produced in the United States that a chapter devoted to the details of its manufacture is really of no practical value. . . ." (Davis, p. 95). Except for a gaggle of handmade paper mills in various countries, including the United States and Canada, Mr. Davis' statement still stands—though it appears to be threatened by the present rediscovery of paper and pulp as mediums for artists.

Current economic realities have forced the small number of mills in operation in North America to price their handmade papers at high levels. In defense of present-day costs of handmade paper, here are some observations by Henry Morris, papermaker, printer, publisher, and sole proprietor of the well known and highly esteemed Bird & Bull Press: "Actually, the New York prices of English handmade paper, as late as 1968, were too low to provide a fair return to the mills. I did a little research on the subject of prices and was surprised to learn that when adjusted to real purchasing power, the price of a ream of 20 × 25 inch handmade paper was actually 30% lower in 1968 than the same ream in 1928. Since 1968 there have been several increases, and using the same yardstick, I believe this paper is now properly priced and includes a sufficient profit for the makers. The question arises as to whether the market will support these prices which, although high, are necessary and reasonable under present-day conditions." (Morris, "Letter to the Editor." *Fine Print,* Vol. 2, No.2. April, 1976, p. 20.)

Given a choice, professional artists prefer handmade paper to all substitutes. A visit to any historical collection of prints and drawings will bear eloquent witness to the previous statement.

So. Buy handmade paper.

But, if you discover that the purchase of handmade paper is beyond your means, what, if anything, can you do about it?

DOING IT YOURSELF

As in other areas of life, there are those who believe that producing paper for their own artistic purposes would be so time-consuming, so expensive, so enervating, so super-specialized, as to preclude the primal concerns of the artist.

Simon Barcham Green, proprietor of the Hayle Mill in England and "a ninth generation papermaker with a degree in paper science" offers this sober judgment:

". . . a professional outlook is essential. The handmade paper mill is a small craft *industry*. It serves other craftsmen by providing an important material. There is room for artistic papermaking—but not a great deal. So beware of the arty-crafty approach; it should not be allowed to obscure the true role of handmade paper: that of a means and *not* an end in itself. This view will not coincide with that of some good friends making paper by hand in the U.S.A. It is a personal one, but sincerely held, and I hope the implications will cause no offense." (Green, "Making Paper by Hand at the Hayle Mill in England." *Fine Print,* Vol. 2, No. 2. April, 1976, p. 18.)

This attitude (which I respect) if held by all, would have long since removed certain basic, artistic freedoms from humanity, including the freedom to fall flat on your face when trying something just beyond your reach. It could also have removed the freedom to flounder, and the freedom to succeed. It may seem, at first blush, that I am trying to reinvent the wheel, to fly in the face of generations of know-how, super-professionalism, technology, and the forward march of the machine. Yet, I firmly believe that an indescribable satisfaction will be yours, can be yours, if you master the simple (of course, it is complex) process of making handmade paper.

Numbers of us, and our little band grows daily in geometric leaps, prefer to manufacture our own materials—when we can—for diverse reasons, including the obvious one of financial savings. We delight in pitting our human resources against, and combining them with, those offered by nature and man to produce that which will serve our needs and standards to meet our various goals. The pleasures derived from making, building, and preparing your equipment, tools, materials, and supplies are truly satisfying.

A few hand tools on part of a wall in The Bird & Bull Press. Photo the author.

JOYS AND FRUSTRATIONS IN MAKING PAPER

If you feel at home with basic tools and materials; if you can follow certain recipes for papermaking knowing, in advance, that *your* paper will be unique—for reasons that will unfold as we proceed; if it warms your heart to know that you can save yourself many hundreds of dollars by recycling pure rag scraps of paper and matboard, worn-out clothing, and ancient linen tablecloths, then you may, in time, learn to savor the intoxication of pleasure found in papermaking.

If you wish to emulate Rembrandt's experimental attitudes toward handmade paper (his dissatisfaction with the papers produced by seventeenth-century Dutch paper mills during his lifetime prompted him to buy German, Swiss, Japanese and French papers, and some from East India, as well as the specially treated skins of calves, kids, and lambs, called vellum), you will have to print the same etching, as did he, on many kinds, qualities, and colors of paper to search out and find an appropriate marriage between your image and its support.

Inevitably, you will spoil some sheets. Some of you will spoil many. Solving certain problems will create others. But, that is the way one learns to make paper or space ships. Frustration will not be your lot, unless you seek the perfection of a professional maker of handmade paper only hours after reading this or any other book. Obviously, as with most good things in life, including making love or wine, it takes time, grace, sympathy, and a little tender loving care.

A DELIGHTFUL, INCURABLE DISEASE

Meet a papermaker and you meet a most peculiar person: he or she is bright of eye, clean of hand, and beats neither spouse nor children. Papermakers just beat rags and grow things.

Papermakers are a happy lot, a special group in love with what it

Henry Morris in a rare moment of rest. Photo the author.

does, sharing an incurable, contagious, low-budget, high-fevered, mysterious disease with all friends and acquaintances: papermaking.

What are the symptoms of this epidemic disease? Apocryphal writings have long suggested that a positive correlation existed between the quantity of alcohol an individual consumes and the quality of the paper said individual produces. But even if one is addicted solely to carbonated soft drinks, coffee, or tea, the primary characteristic of the disease is an unquenchable desire to make one sheet of paper, or a work of paper, after another ad infinitum; other evidence includes a sensual pleasure in wearing cotton and linen costumes which, at a moment's notice, are ripped off one's back and transmuted to handmade paper. A collector who sorts and labels rags for future use may be suspect; others who admit, albeit reluctantly, that they grow a few acres of flax on the back forty, who dye the flax and put it through a flax break, scutching knife, and board, who whip and comb it through a flax heckle before beating it and turning it into paper are, obviously, hooked for life. People who use a colorful jargon composed of simple, earthy words may be papermakers aborning, especially if the words have meanings other than those shown in standard dictionaries.

Individually, papermakers are male and female artists who can be wedded to either personalized blue jeans or generalized banker's or engineer's gray flannel; who either wear no socks or sport the latest fad in leg coverings; who are bemoccasined, booted, or wear made-to-order shoes made in London; who may seem scruffy looking as befits a certain age group or pass as examples of walking dress mannequins; who are technically oriented people dependent on precise equations or free spirits whose attitude to paper formulas and papermaking (and all else) in general is summed up in hunch or intuition—a pinch of this, a smidgin of that. But, it works.

By and large, they are an interesting group, a cross section of all humanity.

EFFECTS OF VARIOUS PAPERS ON YOUR WORK

As mentioned earlier, the specific support on which a drawing, print, or watercolor is made produces a different visual phenomenon than the drawing, print, or painting would make if designed on any *other* kind or quality of paper.

When considering three-dimensional treatments of paper (when paper is treated as a medium in its own right) the nature of the materials used in making paper, the length of beating time, and other myriad factors enter into complex relationships that guarantee the uniqueness of our own handmade paper. These factors cannot but help influence artistic expression to some degree as we, in turn, bend the handmade paper to our own esthetic ends.

The relationship between figure and ground, between image and its support, between the design and the particular paper cannot but engage the eye actively, as it contemplates the composition.

Thus, the same image on a smooth, shiny, bright paper; on white, off-white, gray, or other colored paper; on heavily textured, coarse, tough paper; or on a highly

(Above) Albrecht Dürer. The Virgin with the Child in Swaddling Clothes, 1520. Engraving. 5⅝ x 3¾ inches. Courtesy University Art Collections, Arizona State University, Gift of Mr. and Mrs. Orme Lewis. (Right) Detail of Dürer engraving enlarged 50 times. Photo Cathleen Antonie.

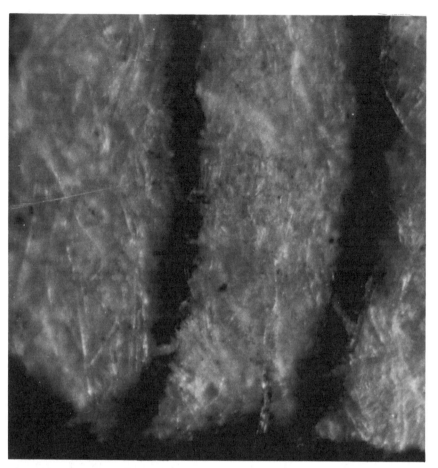

(Above) John Marin. Barges, 1905. Pen and ink drawing. 10 x 7½ inches. Courtesy University Art Collections, Arizona State University, American Art Heritage Fund Purchase. (Right) Detail of Marin drawing enlarged 50 times. Photo Cathleen Antonie.

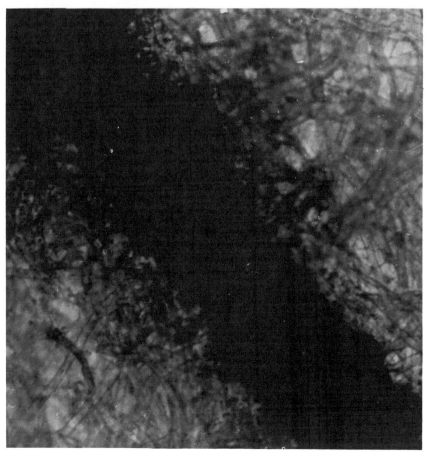

personalized sheet of your own handmade paper designed to fit that image is a different total image.

QUALITIES OF A GOOD PAPER: PARADOXES

At this juncture, we are in difficulty —especially if we insist on honesty and objectivity. We are concerned with more than the craft of making paper; we view papermaking, by artists and for artists, as more than a means to an end—rather, as an end in itself and/or a highly artistic craft.

To quote Ruskin, "Nothing is a great work of art for the production of which either rules or models can be given. It is not an art but a man-ufacture." We must offer rules or models; many of you wish to use handmade paper for drawing, printmaking, watercolor painting, and sculpture, and each of you viewing handmade paper in a dif-ferent manner has certain sets of requirements. The following para-graphs, therefore, reveal a series of paradoxes just as do the contri-butions of practitioners.

Good handmade paper is strong, durable (yet it can also be ephemeral); good handmade pa-per should shrink uniformly in all di-rections (but some artists will woo distortion and distress their papers deliberately); *ghp* (forgive the shorthand here) should lie flat (yet certain artists will revel in making it do otherwise to meet a special need); and *ghp* should be opaque, if it is to be printed or painted on (and even as this is being written I *know* that there are men and women who demand the opposite). In addition, *ghp* holds a watermark well because of its manufacture— the pulp lies over the raised "mark" throughout the making of the sheet providing a translucent image, whereas machine-made paper re-ceives a watermark when a dandy-roll is pressed into the wet pulp, compressing it and darkening the image, after the sheet is formed; yet, I know without being aware of his or her name that someone,

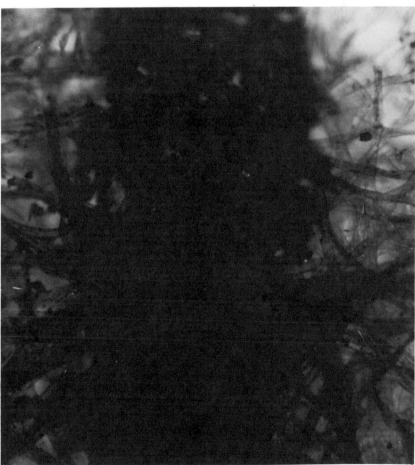

(Top) John Marin. White Mountain Country, New Hampshire, *1927. Watercolor. 20 x 27½ inches. Courtesy University Art Collections, Arizona State University, Gift of Oliver B. James. (Above) Detail of Marin watercolor enlarged 50 times. Photo Cathleen Antonie.*

somewhere, will use the latter technique, sometime, in forming handmade paper—just because. In general, *ghp* should be receptive to ink or paint and should allow these mediums to dry (I am also certain that someone will create papers that will do otherwise for valid artistic reasons); *ghp* should reveal a difficult-to-describe texture, surface, and character—qualities that one senses or feels without even touching the sheet (yet, we cannot rule out the possibilities that opposite properties will be sought and integrated into a new whole by some unknown individual).

Thus, paper may be fragile or tough, ephemeral or durable, very soft or very hard, transparent, translucent, opaque, absorbent, abrasive; it may form the walls of houses and resist fire or be the basis for furniture in all the rooms of an apartment or house; we *know* it is long lasting (500 years or so in the western world and about 2,000 or so in the east); it can be teabag-thin or as thick as an adobe brick; it may serve as scented, personal stationery or as a monumental, rigid piece of sculpture, in which the medium is an integral part of the message.

UNIQUENESS OF YOUR OWN PAPER

We are no longer surprised when a dozen professional artists, working from the same model or landscape, produce a dozen different works of art. It is taken for granted that *your* chowder (I base this upon a careful reading of Mrs. Murphy's famous recipe) will taste unlike that of anyone else. We recognize, if we travel, the enormous differences in color, bouquet, and taste of the white and red wines offered in every village we visit.

Is there a standard taste for spaghetti sauce? Is there but one recipe for curry? Do you know of only one way to get to Heaven? What is a poem? How does one dance the role of *Giselle*? Is the unwritten novel running 'round your head the same as mine? How do you produce a first-rate film? Compose meaningful music? Play a jazz piano? Stalk an Arctic char (a fish)?

The act of making handmade paper is not unlike the reactions of the proverbial blind men reacting to an elephant. No matter how closely you follow the recipes contained in this book, no matter how precise your measurements, no matter how carefully you control all of the elements in this capricious process, your paper will not be precisely the same as anyone else's. There are so many factors involved in this simple but complex, idiosyncratic phenomenon, that the number of permutations and combinations possible in this exercise of esthetic freedom are infinite.

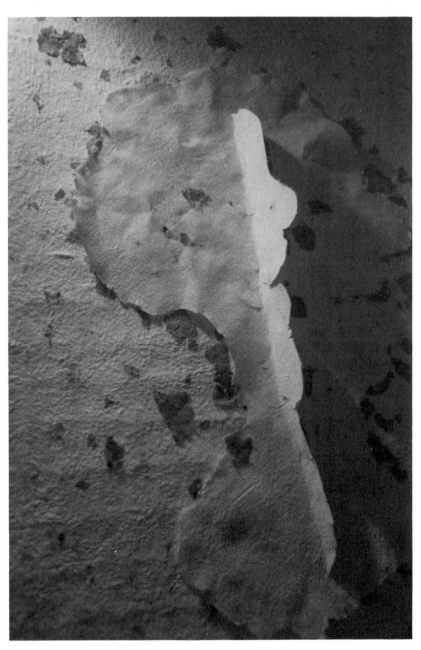

Jules Heller. Detail from Head, *1975. Handmade cotton, hemp, torn and folded. 30 x 20 inches. Collection Dr. and Mrs. David Holbrook, Toronto, Canada.*

PART I. PRACTICE AND SOME THEORY

1. WHAT IS PAPER, ANYWAY?

Rags are as beauties which concealed lie,
But when in paper, how it charms the eye!
Pray save your rags, new beauties to discover,
For of paper, truly, every one's a lover;
By the pen and press such knowledge is displayed
As wouldn't exist if paper was not made.
Wisdom of things, mysterious, divine,
Illustriously doth on paper shine.

—Anonymous, *Boston News Letter,* 1769

At right is a portrait of the first papermaker. The first papermaker has prevailed on this earth for 300 million years or so, since the middle Coal Age. She belongs to the order *Hymenoptera* of the family of *Vespidae* and the super-family, *Vespoidea;* she is one of more than four million species of insects that buzz, whir, sting, dig, glow, bite, do unusual things, or go bump in the night.

She is a paper wasp who has been engaged in the art of making paper for millions of years; a winged queen of a complex social order who macerates dry wood in her mouth and employs the pulpy result to create a habitat of paper for her empire; her paper structure, or nest, strikes envy in the hearts of contemporary architects and fear in the limbs of small boys and girls.

DEFINITIONS OF PAPER

Compared to the arts of pottery, glassblowing, weaving, engraving, calligraphy, and printing, the art and science of papermaking has been little recognized by large audiences. The need to express an idea, proclaim the law of the land, depict and clarify a fuzzy image running 'round one's mind, write a love letter or a declaration of war— all of these goals and more require a support to carry the message (Marshall McLuhan's aphorisms notwithstanding).

Such messages describing events and portraying ideas were carried on stone walls, clay tablets ("And you, O son of man, take a brick and lay it before you, and portray upon it a city, even Jerusalem. . . ," Ezekiel IV.1.), wax, lead ("Oh that with an iron pen and lead they (my words) were graven in the rock forever," Job XIX.24), ivory tablets, palm leaves, the skins of animals and the inner barks of various plants and trees. The messages were written with wax, plaster, or chalk worked with a stylus on wood; they were borne on fish skins, snake skins, and the shells of oysters and tortoises. Finally, they were carried on sheets of

A wasp, the first papermaker.

cotton and linen.

It is believed that cotton (called *carbasus* in the ancient world)— and its many uses—was first appreciated in Abyssinia, India, Senegal, and the Sudan. It was introduced to Assyria about 700 B.C. by King Sennacherib. It is a member of the mallow family and its deep, red-purple blossoms produce bolls of silky cotton about which Pliny noted, ". . . in the island of Tylos are trees that bear wool!" (Pliny, Natural History, XII, 21.)

Lost in the detritus of the Stone Age, flax was another of the crops that our ancestors cultivated from time to time, dependent on conditions that on occasion were quite well beyond their control. The Goddess Isis was worshipped as the inventor of this remarkable material, which, when processed, became the fine white linen still associated with purity in some societies.

Even this brief description of the supports on which persons communicated thoughts to one another strongly suggests possible origins of the words biblos, codex, leaf, library, papyrus, charta, and so on —all referring, in some manner, to the materials or the process of the activity of making paper.

My 16-pound Merriam-Webster defines paper as, "A substance made in the form of thin sheets or leaves from rags, straw, bark, wood or other fibrous material, for various uses." More precisely, paper may be made from pounded, bruised, or shredded cellulose-fibered material, including linen or cotton rags, straw, bark, wood, and

almost all the living plants on this good earth.

Before proceeding further, let us attempt to define the nature of vegetable fiber structure. The weight of a single fiber 9000 meters long would be 3 grams! By and large, vegetable fibers are composed of so-called plant cellulose, which appears to contain small amounts of many other substances: holocelluloses (composed of true cellulose and hemicellulose), fats, gums, lignins, mineral matter, mucilages, pectin substances, protein residues, starch, and other organic compounds too complex to attempt description. Besides, I doubt whether my knowledge of chemistry and related disciplines is up to the task. Suffice to say that lignin is detrimental to our purposes; hemicelluloses are somewhat helpful and cellulose is *essential* in this exercise.

On rereading the last paragraph you may properly ask, "What is cellulose?" Another excursion to that same heavy tome states that the term cellulose is "[a]n inert substance constituting the chief part (cell walls) of ordinary wood, linen, paper, rayon, etc. It is a carbohydrate . . . of the same percentage as starch, and is wholly convertible into glucose by hydrolysis." Before defining these latter two terms, note that cellulose exists in almost pure form in cotton, linen, and the pith of certain plants, and, in a more pure state, in filter paper, cotton wool, and linen rags.It is tasteless, white, odorless, nonvolatile, and is insoluble in water. It may be written as $C_6H_{10}O_5$ and, for those of you who truly wish to pin it down, it has a specific gravity of about 1.45. Cellulose is part of a class of substances called polymers (a union of molecules composed of a number of similar molecules and, in this instance, grouped together end-to-end, to form a long chain).

To return to glucose: you may know it as corn syrup. "It is a light-colored, uncrystallizable syrup (sugar) obtained by the incomplete

Dard Hunter's Particular Definition

Sooner or later, if you are interested in any aspect of paper, Dard Hunter's name is invoked—and properly so. His researches and publications on papermaking and its history and practice crowded the days of his whole life; his contributions to the field have inspired many individuals in more countries to become papermakers. The renaissance of this medium, in North America at least, may be almost wholly attributable to him through a growing chain of individuals. Here is his definition of handmade paper: "To be classed as true paper, the thin sheets must be made from fiber that has been macerated until each individual filament is a separate unit; the fibers are then intermixed with water, and, by the use of a sievelike screen, are lifted from the water in the form of a thin stratum, the water draining through the small openings of the screen leaving a sheet of matted fiber upon the screen's surface. This thin layer of intertwined fiber is paper." (Hunter, *Papermaking,* p. 50). There are some papermakers today whose procedures and methodologies are at variance with those of the Master, but they will be discussed later on.

hydrolysis of starch and containing chiefly maltose, dextrin, and dextrose."

As for the term hydrolysis, it is "[a] chemical process of decomposition involving addition of the elements of water. In many cases it is induced by the presence in small amount of an enzyme, a dilute acid, or other agent."

So much for dictionary definitions that assume the reader is familiar with the subject of chemistry as it relates to papermaking. Curiously, and as Alice said, "The world gets curiouser and curiouser," the chemistry of this very simple process is highly complex and still a subject of controversy among professional chemists.

Paper may also be defined as a felted web, fabric, or tissue of fibrous cellulosic material of uniform thickness, color, strength, and surface. The tissue is composed of felted, individual fibers of cellulose which are beaten (refined and bruised) so the cellulose comprising them may become more or less hydrated (combined or assimilated with water) to form a wet pulp of adequate cohesive quality. This wet pulp becomes a thin, even layer as the result of dipping a mould and deckle (a wire fabric or

screen stretched across a wooden frame with a removable wooden cover) into a vat filled with pulp, giving it the shake or stroke—literally shaking the pulp from right to left and back to front, in four directions, to give strength to the paper-to-be—and then allowing the superfluous water to drain through the wire mould. The wet, even-layered pulp is then submitted to the squeeze (placed in a standing press or hydraulic press under great pressure to drive out the water), and is then further dried and finished in a manner appropriate to the desires of the papermaker.

To summarize, the primary ingredient of paper is cellulose fiber (found in all living plants to a greater or lesser degree); cellulose is a carbohydrate that can be transformed into glucose by hydrolysis. Child's play, wot? Especially, since the materials and methods of making paper are known everywhere, and, with a modicum of instruction, may be realized and enjoyed by children of all ages.

LAID PAPER

Lost in the dustbin of history, the inventor of the laid paper mould deserves, at the very least, a toast from those of us alive and well who

use this person's brilliant invention to make our paper. It is difficult, if not impossible, to define laid paper without reference to laid moulds. Briefly, the mould from which laid paper is made is a hand-crafted masterpiece composed of a varying number of vertical laid wires (brass) or bamboo strips (in the East) to the inch, held rigidly in place by so many chainlines or stitches of very thin wire, horsehair, or the equivalent. Painstaking craftsmanship is evident in the look and feel of an old laid mould. Holding a sheet of laid paper to the light will reveal the strong vertical pattern (true watermarks) on the sheet, as well as the less noticeable chainlines.

WOVE PAPER

Certain authorities believe that John Baskerville (1706–1775), in an effort to obtain a smoother surface for printing, developed (or should we say rediscovered) wove paper. Others suggest the honor should go elsewhere. Held to the light, there should be no regular, linear pattern at all, but rather a sort of regular, evenly mottled appearance valued and enjoyed by many printers, printmakers, and painters.

The mould below right is, perhaps, the oldest kind of paper mould known, at least according to the speculations of Dard Hunter. Wove paper derives from such moulds in which the wove covering across the mould is not unlike that of a household window or sliding-door screen except that it is made of rustproof wire or synthetic fiber-like materials and the mesh is infinitely finer than that used to protect our precious interiors from members of the insect air corps. The moulds illustrated were superbly crafted of Honduras mahogany and rust-free materials.

MOULD-MADE PAPER

Mould-made paper is manufactured on a cylinder-vat or cylinder-mould machine—"water is drained through the wire cloth on the sur-

(Left) A laid mould with deckle in place made by James Yarnell. Note watermark of bunch of grapes.

(Below) At the extreme left, a wove mould of fine-mesh, synthetic fiber woven-screen. The other moulds are all laid moulds of various sizes. The wove mould leaves no laid or chain lines on the paper. Note Yarnell's watermarks.

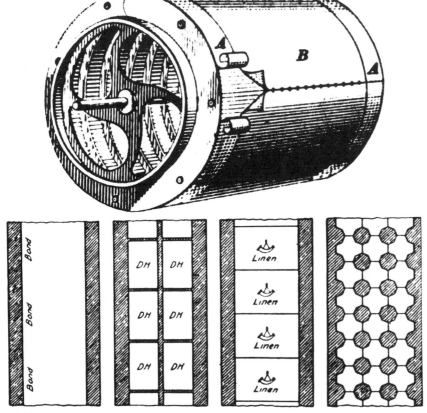

Cylinder from cylinder-vat (mould-made) paper machine and methods of manufacturing multideckled sheets.

The Fourdrinier, or endless-web, machine.

face of a rotating cylinder partly submerged in a vat of fibres suspended in water. The water is sucked away from the inside of the cylinder leaving a mat of fiber on the outer surface and as this emerges at the top of the vat it is transferred to an endless felt. . . .'' (Watermark 74, unpaged.)

Mould-made paper may have two or four deckle edges and appears, to the uninitiated, to be handmade. Close examination of a mould-made sheet, however, will reveal the evenness of the deckles, the lack of "character" of the paper, and the deadly similarity of the individual product to all other sheets of mould-made paper. You would have to place a mould-made sheet and a handmade one side by side, touch them, tear them, and fold them, to really sense the differences.

Mould-made paper can be made in an endless web, which manufacture organizes the fibers in the direction of rotation. This results in paper that is stronger in one direction (the direction of rotation) rather than across the paper.

Waterproof cloths are fitted at the ends of the machine felts to provide two deckles (see the figure above left for a graphic explanation). Mould-made paper can also be made with four deckle edges by laying additional waterproof cloths at specified sheet lengths across the felt of the machine.

ON MACHINE-MADE PAPER

The paper machine was invented by Nicholas-Louis Robert (1761–1828). It is believed that his reason for inventing the machine was to avoid the bickering and contentiousness he found among those who made paper by hand in the mill of Francois Didot (1730–1804) in Essonnes, France, where Robert was employed as inspector of personnel.

Robert truly invented a machine which could make paper in an endless web, as wide as the machine he devised. Strangely, the long web of paper was cut and dried in

the manner of handmade sheets—
hung on ropes of horsehair in drying rooms—since there was no
need then for such long sheets of
paper and no printing press yet developed to utilize such paper. Robert probably did not even conceive
the idea of drying the paper on the
same machine, also mechanically,
so an endless web of paper could
be produced without the aid of a
single human hand. His employer,
Didot, assisted him in every way,
yet, after having secured a patent
for his invention, Robert and Didot
soon were at loggerheads in the
courts.

Didot's brother-in-law, John
Gamble, was proprietor of an English paper mill. Henry and Sealy
Fourdrinier of London were called
in by Gamble and soon developed
a large machine along the lines of
Robert's invention. Then, through
Gamble, Bryan Donkin entered the
scene—he was both a visionary
and a skillful mechanic, who re-
vised and improved upon the paper
machine that, more or less, we
know today. (Except that today
they are enormous, automated
monsters.)

The Fourdriniers, thanks to a flaw
in their original patent, spent a fortune but received no royalties for
their efforts; curiously, we still call
the machine by their surname.

ON PAPYRUS

My faithful Merriam-Webster
source defines papyrus as "[a]
tall sedge (*Cyperus papyrus*), native to Egypt and adjacent countries. . . . Its fiber served . . . as a
writing material by the ancient
Egyptians, Greeks, and Romans. It
was prepared by cutting longitudi-
nal strips, arranging them cross-
wise in two or three layers, soaking them in water, and pressing
them into a homogeneous surface.
The use of papyrus for literary and
commercial manuscripts extended
from the fourth century B.C. or earlier, to the fourth century A.D. with
occasional use until the ninth century." Thus, in the strictest sense of
the term paper, papyrus does not

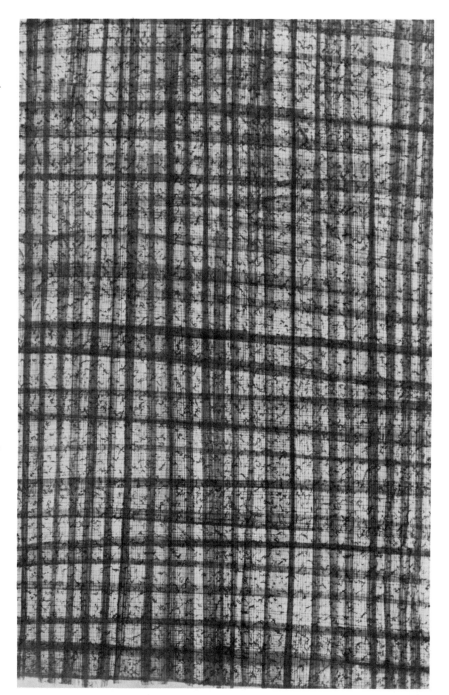

*A sheet of papyrus photographed against the light to reveal its mode of manufac-
ture. Collection the author. Photo Merrylee Stephenson.*

qualify: it is not macerated into separate fibers, nor is it sieved through a screen. However, there may be certain papermakers today who are not inhibited by definitions and who may delight in accommodating a nonpaper to their own expressive means—with salutary results. For the reader who wishes to purchase a sheet of papyrus to investigate and/or use it, please consult the Suppliers List.

ABOUT TAPA

Tapa is manufactured in tropical countries around the world; it is made in Africa, Indonesia, Melanesia, New Guinea, the Malay Peninsula, Polynesia, and South America. Tapa is a kind of bark cloth or bark paper, depending upon its function.

"From museum specimens, it is evident that Polynesian bark cloth is superior, on the whole, to that made in any other area. Within Polynesia itself, though Tahitian tapa was held in high regard by Cook

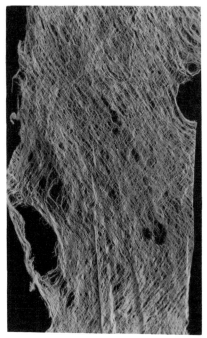

(Top) Fijian tapa purchased in Tahiti. 52½ x 74½ inches. Collection the author. Photo Merrylee Stephenson.

(Above) Samoan mulberry fibers—early stage. Collection Robert Nugent. Photo Merrylee Stephenson.

and other writers, the Hawaiian bark cloth (*kapa*) displays the greatest varieties of texture and colored designs." (Te Rangi Hiroa —Sir Peter H. Buck—p. 166.)

Tapa, obviously, is another nonpaper. Robert Nugent, a California artist, described his experiences in making tapa during a year spent on the primitive island of Tau, in the Manua group of American Samoa in 1971 (see page 148 for work by Nugent): "The material is made from young mulberry trees only a few inches in diameter. A knife is used to slit the bark down one side, and then this outer layer is peeled away. It is the inner portion of the mulberry bark that is scraped away with a shell (*atini*) shaped like a sort of fan. The young paper mulberry tree is called *tutuga* (g is pronounced ng). It is softer and finer than Tongan Paper. Once the fibrous material is separated from the bark with the aid of the shell, it is soaked and beaten daily, elongating and separating the fibers. A piece of material 4 × 6 inches will produce a piece of paper 5 feet square, but with a lot of holes. The mulberry produced in this manner is not really paper by Dard Hunter's definition, since the material has not been macerated, but simply elongated like Egyptian papyrus.

"The Samoans then cover the holes in this material by gluing several layers of the mulberry together using overripe breadfruit as glue. The island I was on is the last place natural dyes are still used for painting (mostly acrylics now): purple from the banana trunk, black from the candlenut, red from the *loa* berry of the lipstick tree, yellow from the roots of the turmeric plant, and from the bark of the *o'a* tree comes brown.

"While I was working and living in the islands, I made small collages of mulberry paper and used natural dyes for pigments. Although I was an oil painter when I arrived, I was a papermaker when I left and have not painted in the traditional sense of the word since Samoa."

Antonio Pigaletta, writing of the Molucca islanders on Magellan's 'round-the-world voyage in 1521, noted this in his journal with regard to tapa: "They make their bark cloth in the following manner; they first soak a piece of bark in water and then beat it with clubs until it assumes the desired length and breadth. After this treatment the bark looks like a silk fabric of very fine texture." (CIBA 33, p. 1175.)

In his book, *My Life With Paper*, Dard Hunter describes in some detail the manner in which 28 women worked on strips of the inner bark of the mulberry tree to produce one enormous sheet of tapa measuring 24 feet by 90 feet!

In Fiji, the indefatigable Hunter noted the following in his diary, "Before arising the following morning I could hear the rhythmical pounding of the *tapa*-beaters, a sound . . . distinguishable from all others." (Hunter, *My Life With Paper,* p. 94.)

Jean Pucelle. The Hours of Jeanne d'Evreux, *XIV century French manuscript. 3½ x 2⁷/₁₆ inches. Grisaille and colors on vellum. Courtesy, Metropolitan Museum of Art, New York.*

Camila Hernandez. Untitled, n.d. Bark paper (amatl) drawing on bark paper. 22¼ x 36½ inches. Collection of Juan Manuel de la Rosa. (This innovative technique was invented by Dona Hernandez (b. 1917?), who lives in San Pablito—a small village high in the Sierra Madre Mountains of Mexico.)

ABOUT RICE PAPER

The tree from which "rice" paper is grown flourishes in the hills of northern Taiwan. It is cut into one foot sections that are about two inches wide. The pith is forced out of the section and placed in bamboo canes to swell and "grow" larger in diameter. When the swollen pith is removed, if an extremely sharp knife is held against it and used with precision, a single sheet of paper up to four feet long may result, on occasion.

Thus, what once was an inexpensive paper found in almost all art supply stores, used by sundry persons for printing woodcuts and other relief prints, and widely known as "rice" paper, is a misnomer in every sense. This "paper" is neither made from rice nor does it meet Dard Hunter's criteria for paper. This smooth white material "is cut spirally from the pith of the *Fatsia papyrifera* [now *Tetrapanax Papyriferum*] and, like papyrus and tapa, is not a macerated fiber material. . . ." (Dard Hunter, *My Life with Paper,* p. 6.)

To further confuse everyone, paper may be made (and is) of rice *straw,* which the Chinese term *ts'ao.*

ABOUT PARCHMENT AND VELLUM

Parchment was probably used more than three thousand years ago; Eumenes II, King of Pergamum (197–159? B.C.) is usually credited with its invention (perfection?), because it is believed he wished to produce a writing material to rival papyrus (then difficult to obtain from Egypt).

Parchment is made from the fleshy side of the split skin of sheep; a strong leather is obtained from the wool or grain side. If the fleshy or lining side of the sheep skin is not suited to the making of parchment, then garments of chamois or suede are made. If the skin is not suited to clothing, the result is a chamois rag.

Vellum derives from still-born or newly born calves or lambs. Both parchment and vellum are created using the following practices today —as they were centuries ago. First, the goat or lamb skin is washed thoroughly. Then it is rubbed with lime, the hair is removed, and the skin is scraped with a curved knife and washed again. After being stretched tightly with leather thongs on to a four-sided wooden frame, the scraping continues until all irregularities are removed. Thus, we witness a skin of even thickness throughout its area. Finally, the skin is dusted with powdered chalk and again rubbed with fine pumice. Since neither parchment nor vellum is tanned, they feel and look more like paper than leather.

ABOUT HUUN AND AMATL

The ethnic groups that inhabited Mexico and Central and South America early on made a kind of paper from the inner bark of the generae *Morus* (mulberry), *Ficus* (fig), and *Cannabis* (hemp). The last mentioned, an herbaceous annual plant (*Cannabis sativa*) is one of man's oldest cultivated nonfood plants, believed to have originated in Asia. Aside from the plant's medical and nonmedical pharmacological uses which are not our concern, it is suggested by authors writing for the *Chemurgic Digest* and/or offering papers at a meeting of the Canadian Institute of Forestry that "a modern hemp industry be encouraged in North America for ecological as well as economic reasons, since cultivated cannabis is several times more efficient in producing pulp for paper on an annual acreage basis than is forest woodland." (Cannabis, pp. 14–15.)

To return from this interesting digression. The Mayans, proud people of the Yucatan, wedded the making of *huun* (the name of their beaten bark substance) to their need for figurative and textural expression to produce vast numbers of individually illustrated manuscripts, documents, and codices (books), both sacred and profane, in hieroglyphic writing.

The Aztecs, who followed on the historical scene upon the decline of the Maya, developed their beaten paper (*amatl*) from the equivalent of a cottage industry to that of a major industrial phenomenon.

Hans Lenz, writing of a visit to the Otomi Indians in Southern Mexico (in San Pablito, the state of Puebla, and the municipality of Pahuatlan) states that "[b]oth sexes procure the raw material for 'paper' making, but its actual manufacture is only done by the women. Several hundred of them are so employed." (Lenz, *Mexican Indian Paper.*) To keep the record straight, it should be pointed out that women manufacture paper in Polynesia, Macronesia, and elsewhere in the East.

The inner bark of moraceous trees (mulberry, fig, and hemp) is stripped in one-inch or so widths, boiled in vats of water for about ten hours with a large quantity of ash, washed, and then placed side by side on a board, each strip slightly overlapping the next. The strips are then beaten with stone tools to become a single sheet of paper and, finally, placed in the sun to dry on the board.

The Otomi use their *amatl* papers for ceremonial purposes, including the making of dolls to call up good (*Xalamatl* Limon) or evil (*Xalamatl* Grande) spirits. It is interesting to note, in passing, that the originators of paper used the same material (paper) for equivalent purposes many centuries before.

2.
A TWO THOUSAND YEAR OLD APPROACH

A paper man will tell you
he turns 'old shirts into new sheets:
and that indeed is what he does;
but a long and toilsome journey lies
between the old shirt and its apotheosis.

—E. Philpotts, p. 11

Making handmade paper is as easy as falling in love (or out of it)—given sufficient natural or manmade material, the proper tools and equipment with which to beat or refine the material into pulp, moulds and deckles with which to form the paper, and other equipment with which to dry the sheets.

Unfortunately, there are certain basic requirements and unalterable steps that must be followed by amateur and professional alike.

Preparing the natural or manmade material from which you make paper.

Beating or refining the material.

Forming the sheets.

Drying the sheets.

See Bibliography: Kunisaki, Jihei.

THE PROCESS OF MAKING PAPER —A PICTURE STORY

Before traveling the "long and toilsome journey," using substitutes and equivalents with which to make handmade paper, I suggest it would be useful to witness the entire process from beginning to end. Following, therefore, is a demonstration in which the proper tools and materials are used by James Yarnell, who has the distinction of owning the only paper mill in Kansas as an outlet for one of his many interests. It will also be useful to read the prose, a most delightful summary on the subject of handmade paper rewritten for this book by Walter Hamady: artist, papermaker, typographer, printer, publisher, professor, and a man who enhances the art and literary environment by his works. All the photographs in this chapter are by James Hellman.

James Yarnell, proprietor of the Oak Park Press and Papermill, builder of his own Hollander (a beater originally designed in the 17th century), and writer of these captions (with certain liberties taken by the author) in the privacy of his home-workshop in Kansas.

(Top) The "Irving E. Lee." (You may note a certain resemblance between the Hollander and the stern-wheelers that once plied the major rivers of this country; you may also take the measure of Mr. Yarnell's effervescent wit.)

(Above) Beater with the chassis raised for cleaning. Everything employed in papermaking must be thoroughly cleaned each time it is used. Note the adjustment bolts on the far end, at the top of the photo, to raise or lower the roll/bedplate relationship.

(Top) Detail of the roll. The blade on the side of the roll (center of photo) prevents build-up of pulp between the roll and the side of the tub.

(Above Left) The bedplate (bottom) of hard aluminum alloy bars conforms to the width of the roll (top). The spacing between the roll and bedplate is one factor that controls the rate of maceration. The two are never in complete contact, but rags are "rubbed" through the narrow opening for six inches on each pass through the system.

(Above) The beater, empty. The motor rotates the roll to move the pulp through the ramp and bedplate. On leaving the bedplate, the pulp drops over the backfall, then around 180° to the left, through the narrowed section; the bottom of the tub gradually slopes upward, around the near end of the midfeather, up the ramp, and through the roll again.

NOTES REGARDING THE ACTIVITY OF PAPERMAKING
by Walter S. Hamady

"I do not like to make paper nor do I like the papers that are available on the market today. Quite frankly, I am hooked on handmade. Knowing that this is not the time or place to talk of addiction (its history &c.), I will try to give a few ideas concerning handmade paper and briefly sketch the procedure in its making. To preface the preface, I am assuming that everyone knows the position that paper holds in relation to the graphic arts: that paper is an entirely intriguing and fascinating substructure in itself.

"Papermaking is essentially a craft. It can be a highly artistic craft, of course, when an artist is making paper. It is a means to an end and can be exquisitely beautiful. It can play an essential role in the total expression of the graphic arts—though, as it is taken so much for granted, most often it does not. Truly, fine papermaking is many things or it is only high-class Kleenex, depending on one's interest and sensitivity to an absolute and critical element.

"Fine papermaking, by hand, need not pose any ominous-mystifying-mumbo-jumbo problem beyond securing the necessary equipment. The manufacture of handmade paper only requires dexterous love and strong sacrospinalis. A great American papermaker is said to have said that 'any damn fool can make paper.' I take his word for it, because here I am having made some myself.

"The main ingredient for all natural paper is simply *cellulose fiber*. All living plants are made up of this fiber and, properly prepared, can produce some kind of paper. A trip to the dictionary will show that cellulose is a carbohydrate convertable into glucose (an uncrystallizable sirup) by hydrolysis, a chemical process of decomposition involving the element of water. Simple.

(Top) Clean cotton rags cut into bite-size pieces for pulping.

(Above) Roll-hood secured for operation.

"At this point it can be concluded that paper materials—fiber and water—are abundant, inexpensive, and readily at hand.

"It might be interesting to note a few of the many fibers in use around the world: from bizarre-sounding frijolillo, itiki bouroballi, ranbhendi, wallaba, and wanasora wiswiskawlie; to fun-sounding fiddlewood, crowsilk, licorice roots, peppermint, sassafras, and swampbay; to more familiar-sounding barley, milkweed, spruce, wheat, pine, and oak. I would imagine that a lack of natural source material for papermaking in any livable latitude is nonexistent, since any plant that grows is certainly eligible—the only qualifying condition is the quantity of cellulose fiber in relation to the quality of the paper; the greater the better.

"The best paper is made from *used* linen and cotton rags—the highest in cellulose fiber content and are preconditioned to hydrolysis. It seems a prognostication, almost, that the best papers should be made from used rags that human use assisted in preparing.

"The actual conversion from rag to pulp is accomplished by literally *beating* the fibers in water. The principle is as simple as the mortar and pestle, and, if someone wanted to be a rabid purist, he could manufacture pulp in this primitive fashion. But since we are influenced by certain aspects of the industrial revolution, we naturally have a machine called, of all things, a *beater*. The object in making good pulp is to have a variation in fiber length and adequate strength. These objectives are achieved by careful regulation of the beater, as a machine, and by careful, slow feeding of the rag. In short, the making of the pulp is as critical as any other step in papermaking, for the relationship of pulp to paper is like the relationship of paper to printing.

(Top) Beater running with water only, just before adding the cut-up rags. The millrace is narrowed at the far end to speed the flow and prevent eddies or backflow in the "backstretch."

(Above) The Irving E. Lee under a full head of steam, frothing, roaring, and splashing merrily away. (The average beating time from rags to pulp is five hours.)

Beater seen from above. Time exposure shows movement of the pulp as the roll spins to create a kind of "pumping" action. (The movement is counter-clockwise.)

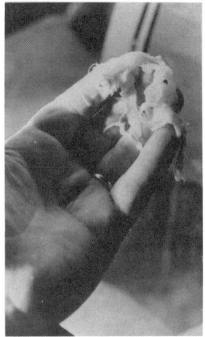

(Above) After being beaten for 30 minutes, the cotton rags begin to show signs of breaking up.

The beater roll is adjusted to the desired clearance with the bedplate. Based upon the desired end-product, adjustments are made periodically; the length of time pulp is beaten also plays a great role in determining the final product.

39

"The time that the beating takes to produce pulp depends on the *muscle* of the rag being used. Progress is checked by pinching a small amount of pulp from the beater, putting it into a small jar of clean water, shaking it vigorously and holding it against the light. This is called a *freeness test* and is performed to see if there is any weave or thread remaining. Another empirical method of determining the state of the pulp is by feeling for *hydration.* When the fibers break down into pulp they hydrate, which means that they become supersaturated with water and somehow give off their own bonding agent called mucilage (some fibers will not do this if they have been treated). Hydration is indicated to the touch by feeling slippery and slimy. At this point the pulp is ready. The pulp to water ratio will be too thick directly from the beater and must be diluted for *forming* into sheets of paper. This is done by adding the pulp, a small amount at a time, to the water in the forming *vat.* This step is called *mixing the slurry.* Now we are ready to form.

(Top) After one hour, the rags have lost their identity, yet threads still make the mass cohere.

(Above) The beater is shut down with almost-finished pulp. Two pounds of rags in 12 gallons of water for four hours has converted several ex-$15 shirts into a tub of pulp easily worth 95 cents. Progress!

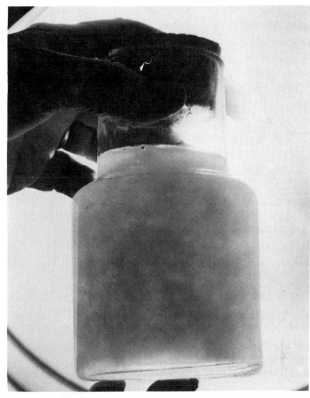

(Above Left) Pulp is tested for consistency by dropping a wad into a jar and shaking vigorously. If it becomes even and milky and has no lumps, visible threads, or particles, it is ready for the vat.

(Above) The milky appearance of this sample indicates that this rag pulp is ready for the vat.

(Left) After four plus hours of beating, the finished pulp is added to a vat full of clean, fresh water. (The thickness of the sheets in a given batch of paper is governed by the proportion of pulp to water.)

"The actual forming of pulp into a sheet of paper is done with a *mould* and *deckle*. The mould is a simple affair. It is a woven metal wire screen attached to a wooden frame. Ribs span the short dimension of the frame to reinforce and prevent the screen from sagging. The deckle is another, slightly larger, frame that fits over the mould. It is removable, and its function is to serve as a fence or wall around the surface of the screen, preventing the pulp from spilling off the surface of the mould. There are countless combinations of different mould and deckle shapes and screen weaves for any dreamable purpose one might have.

"With the deckle securely in place, and with both hands gripping the short sides, the mould is held perpendicular to the surface, then quickly but evenly dipped into the vat, tilted horizontally, and pulled straight up. It is important that this be accomplished in one even continuous movement. While the water is draining, the mould is given the shake, a gentle but firm movement from side to side then back and forth, to unline the fibers, to help the settling fibers weave, and to even the surface.

"Hydration of the pulp is important at this point, as an unhydrated pulp will not hold water long enough to give it the shake.

One of the most "stirring" scenes in the papermaking process. Pulp has a pesky tendency to settle and has to be stirred often. This picture also shows the vatman, the vat, the assboard (drainboard), and the ass (also known as the horn or stay). This may be the fanciest ass in the paper world, partly due to the fact that the proprietor and vatman is also a violinmaker of sorts.

First position of the mould in the sheet-forming process. The deckle frame is held tightly to the mould by hand pressure.

The mould is plunged into the vat, but is not completely submerged. When you do this, bring the bottom edge of the mould back to your body and hold it level.

The mould with a new charge of wet pulp. A rocking action evens the thickness of the pulp, and then a shaking from right to left and back to front causes the fibers to close.

A newborn, soaking-wet sheet on the assboard. At this stage, so say the experts, the product is less than 10% fiber and more than 90% water.

A vatman from Central Madras
Had a simply magnificent ass
Not rounded and pink
As you'd naturally think
But was teakwood, embellished with brass.

—composed by James Yarnell during the meeting
of the First National Papermakers Conference,
Appleton, Wisconsin, November, 1975.

"When the draining is sufficiently completed the deckle is carefully removed. At this stage, the newly formed sheet, called the *waterleaf,* is the most susceptible to damage, and even a drop of water falling from the deckle can cause a thin spot to result in the finished paper. If the sheet should be damaged or malformed in any way, it can be easily removed from the mould by kissing off—accomplished by turning the mould face down and touching it to the surface of the water in the vat. Kissing off cleans the mould of the newly formed sheet and returns the pulp, with a little agitation, to service again.

"The waterleaf is removed from the mould and transferred to a *felt* by a method called *couching* (rhymes with smooching). The felt is first dampened, and then laid on a thin piece of sheet metal on a thick piece of plywood the same size as the felts. This facilitates moving the ensuing stack. The actual couching is done by holding the mould vertically at the side of the felt and with both hands, in a continuous movement, pressing the face of the mould on to the felt. . . . [This] allows pressure contact to move in an even band across the face of the mould, facilitating a relatively easy transfer of waterleaf to felt. This process is repeated by *throwing* another felt and couching again. The paper will only adhere to the felts, not bond with them, so the process can be repeated just as long as there is a supply of pulp and felts. The resulting pile of alternating felts and waterleaves, called a *post,* is then transferred for pressing. This can be done in a variety of ways, all to the same end of removing as much water as possible (the squeeze). I find that the best paper is that which has undergone a lot of pressure, slowly brought to bear, for a good length of time.

(Top) Removing the deckle frame after forming the sheet.

(Above) Mould with the newly formed sheet. Note the deckle in the right foreground. The deckled edges will result when the water is squeezed from the sheet in the pressing operation.

Mould propped against the ass for draining just prior to couching (transferring) the wet sheet to a dampened felt.

Edge of the mould showing the thickness of the wet pulp just after the new sheet was formed and the deckle removed.

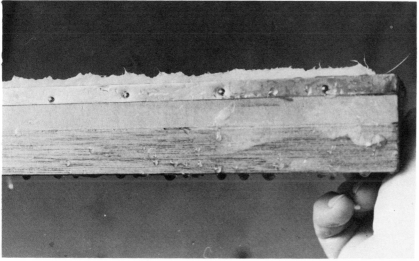

"After removing the post from the *squeeze,* each sheet of still damp paper must be carefully peeled away from the felt and laid out to dry on a large screen called the *loft.* It seems the faster the paper dries the stronger it becomes; by curing (letting the paper age), more strength is additionally imparted. At this point the paper can be further finished by sizing and smoothing; but I am not properly informed to discuss these matters for I am not interested in performing such unnatural acts on the pristine beauty of handmade paper.

"In defense of my unscholarly description of papermaking, which surely horrifies all my puristic peers, I can only say that an article of such brevity affords me not enough room to list, let alone explain, the subtleties and refinements of the process. Truly it is this aspect of papermaking that has all the fragrance of poetry."

(Top) In a very "showbiz" gesture, the proprietor confidently inverts a mould with a newly formed wet sheet clinging to the bottom surface. This probably proves something quite profound, but the proprietor has not yet found how he can capitalize upon it. Actually, the cohesion is what makes couching possible—which is something. . . .

(Above) A pile of felts and a couching table (on the left). The felts are heavy, tightly woven wool, obtained from commercial paper mills. Note the slightly convex surface of the couching table; it is so designed to facilitate transfer.

(Above Left) First position in couching a newly formed wet sheet to felt. A rolling action will transfer the sheet from the surface of the mould to the dampened surface of the felt.

(Above) Finish of couching procedure. The wet sheet has been transferred from the mould wires to the dampened felt.

(Left) New wet sheet after couching.

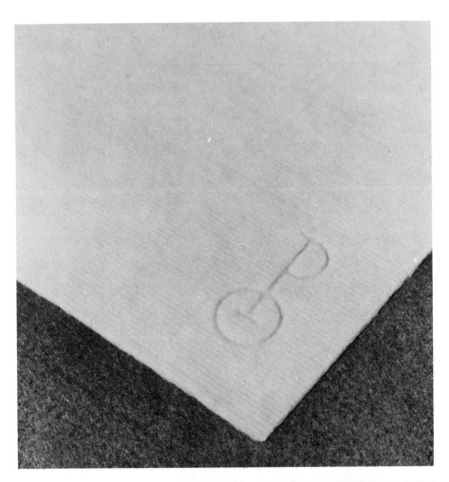

Close-up of newly couched wet sheet showing watermark and texture of the mould wires in what is called a laid mould. (A wove mould would resemble a finer version of ordinary household window screening.)

After couching, another felt is placed on top to receive the next sheet. It is couched and another felt is placed on top of that one, and the process is repeated until a sufficient quantity of sheets (a post) sandwiched between felts results. In the past, a post was made up of six quires (a quire contains 24 sheets).

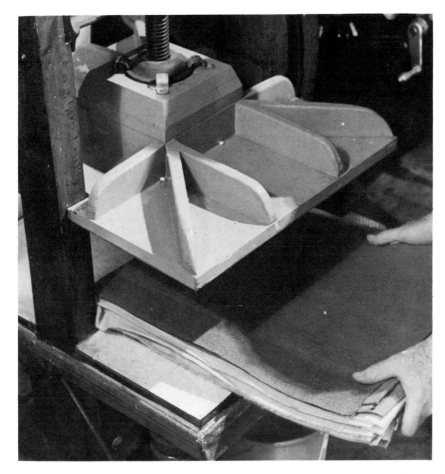

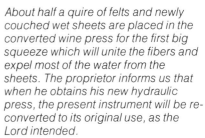

About half a quire of felts and newly couched wet sheets are placed in the converted wine press for the first big squeeze which will unite the fibers and expel most of the water from the sheets. The proprietor informs us that when he obtains his new hydraulic press, the present instrument will be re-converted to its original use, as the Lord intended.

After the first pressing, each sheet is stripped from its felt to be transferred to another dry felt (foreground) for a second pressing in another quire or so of damp sheets. Upon completion of this second pressing, the sheets are taken from the stack of felts, dried, and either tub-sized and calendered or retained as waterleaf (unsized paper).

3. THE SIMPLEST WAY TO MAKE PAPER

*For purity of air and water, chemicals and
working hands is a vital matter to the paper maker.
Every operation must needs be as cleanly
as sleepless precaution can make it.*

—E. Phillpotts, p. 8

*Shall quips and sentences and these paper bullets
of the brain save a man from the career of his humour?*

—Shakespeare, *Much Ado About Nothing*, Act II, Scene 3, Line 260

Coincident with the renaissance of handmade papermaking, a number of individuals, associations, and institutions each claim to offer the *sole* road to heaven (or the other place) with regard to inexpensive and simple approaches.

Given a low budget, a minimal amount of equipment, and a bewildering number of approaches to handmade paper, I intend to provide you with a cross section, a mosaic of offerings from a fair sampling of these several expert approaches within this book.

Now that you have witnessed Mr. Yarnell's procedures, one or more of you may react thusly:

"So!" you exclaim, with delusions of adequacy in your voice, "You must be joking. What is so difficult about that?"

And there, unfortunately, is the rub. This very simple, this apparently facile method of making paper is more or less as easy as

learning to play the violin as well as Isaac Stern, playing ten under par on the most difficult professional golf course in your area when you know you are a duffer, or walking safely across a field booby-trapped with thousands of cultural land mines. As is true in all of the arts, the solution of one part of a problem brings about a host of others much more complicated than the one just encountered. But, that is why artists become artists . . . and . . . perhaps we had better return to our present problem.

May I assume that this is your first experience with the process of making paper? May I assume you do not wish to invest a large sum of money in this enterprise? May I further assume that you would like to use the product of your labor within a reasonable period of time?

The outcome of your first attempt at papermaking may, perhaps, be worthy of accepting a major work

Here are some of the materials and equipment assembled. Note the already mixed batch of pulp (on the right) and soaking, cut linters (also on the right), which will soon be in the blender and subjected to short bursts of speed (about 45 seconds to a minute or so) to produce another batch of pulp.

of art on its surface, a laundry list (shades of Alois Senefelder and the invention of lithography!), or a plastic idea in and of itself. Whatever the outcome, it will unmask the procedure for making pulp and sheets of paper with simple tools and materials, without further ado.

It, hopefully, will whet your appetite for making paper and for investigating the history, mystery, and nature of this fascinating material. And, finally, it will stimulate you to find your own ways to manufacture and employ paper to meet your private, personal, and esthetic needs.

May I presuppose you have a kitchen, a basement, a garage, or an oversize closet—a place to work—either inside or outside your abode with access to clean (pH neutral or thereabouts and iron-free) water and electricity?

PREPARING THE MATERIALS

There are a number of simplistic approaches, kitchen-type variations on papermaking (see Chapter 10) utilizing facial tissues, pH neutral photographic blotters, filter papers, toilet paper, and so on. I find it difficult to believe that very many individuals will want to draw, paint, or make prints on such recycled materials, even though they may be high in cellulose content and pH neutral; somehow these materials leave much to be desired with regard to quality.

For those of you who are champing at the bit, eager to make your first sheet of handmade paper using the "easiest" method available and unhappy with the many words that have intervened between desire and end product, proceed to gather the following:

1. Cotton linters cut up into 1 to 2 inch or so rectangles or, if you truly

(Top)The second batch of pulp being poured into the deckle. Note the consistency of the pulp.

(Right) Reverting to a childhood game, play "patty-cake, patty-cake" to make certain you have an even (or fairly even) layer of pulp.

cannot wait to order linters, pH neutral photographic blotters, filter paper, or even scraps of good quality mould-made papers, such as Arches, Rives, etc.

2. A high-speed food blender.

3. A large sponge (for washing automobiles).

4. A 4-cup Pyrex measuring cup.

5. A plastic vessel of similar size.

6. A wooden frame or deckle of 1 × 2 inch hardwood stock—the size of the paper you wish to make. If you wish to use this method again and again, polyurethane or water-proof the deckle.

7. A package of cheesecloth.

8. One gunny sack or some equivalent loose-woven material.

9. Two pieces of sized canvas larger than your proposed sheet of paper.

10. A large sheet of plate glass, plastic, or some type of hard-surfaced, waterproof tabletop.

11. An old-fashioned clothes wringer, an etching press, or an old bookbinder's press—if you don't own one, try to find a friend or acquaintance who does. If necessary, you can always use a household iron, or just a good sunny day.

PROCEDURE

1. Tear or cut your material (linters, blotters, etc.) into 1 to 2 inch rectangles. (One linter, about 24 × 36 inches, to about 5 gallons of water will provide the concentrated pulp needed for this method.)

2. Add the cut or torn material to the plastic vessel filled with water. Drop the pieces in, *one at a time,* allowing them to soak well.

(Top) Wait for several or more minutes, and then lift the deckle carefully from the pulp. Note how thick and firm and uneven it is at this stage.

(Right) Place another wet piece of cheesecloth on top of the virgin-white pulp.

3. Mix a batch of pulp from the pre-soaked linters or their equivalent by placing a small handful of them in the blender (which should be about ⅔ filled with pH neutral water). Subject the material to short bursts of speed. Continue until you have pulp. With this approach, the fineness of the pulp, its freeness, and so on, are not a critical factor.

4. Wet a piece of cheesecloth (larger than your intended sheet of paper) in water.

5. Spread it out flat on the glass or tabletop.

6. Place the deckle on the top of the cheesecloth.

7. Quickly pour the just-mixed Pyrex container of pulp on to the gauze along one-half of the deckle. Prepare another batch of pulp the same way and pour it on.

8. Prepare a third batch of pulp in the same way, in case it is needed.

When the sheet is bone-dry, you will have produced what is termed waterleaf, or unsized, handmade paper.

Many papermakers will stop at this juncture, satisfied with their end-product and will store it for future use for relief printmaking, printing, typing, or writing letters with a ballpoint pen; others may wish to engage in activities that demand further treatment of the paper, such as painting, lithography, etching, and calligraphy. If so, you

(Top) Starting from the perimeter, work around the sheet with a dampened sponge to felt the sheet (press the water out of the paper). Squeeze the sponge dry when necessary and continue to felt the paper—gently, at first.

(Center) Using all the strength you can muster, move closer and closer to the center of the sheet and continue to felt and squeeze dry your sponge. This could be termed "hard felting."

(Right) When you believe you have extracted all the water from the sheet, remove the cheesecloth from the surface.

must size the paper (see pages 80–81).

FURTHER CONSIDERATIONS

Drying a sheet or sheets of paper is a troublesome activity—you will find other references to this problem throughout the book. Suffice to say that you can allow the sun to work for you, or you can use an infrared heat lamp, an electric household iron, a rope to hang sheets of paper from in an airy loft, or blotters or cloth to make "sandwiches" with paper in between.

With regard to the excess pulp you have made in the manner outlined, you may wish to squeeze it into fairly dry fistfuls, wrap each one in a plastic bag, and store them in a refrigerator until, once again, you wish to make paper. At that later date, break up each fistful of pulp into smaller components and submit them to the blender treatment. (One of my daughters pulped old cotton sheets in a kitchen blender using 1 inch squares and a great deal of water —almost 97 % water to 3 % of matter. She stored the unused pulp in the family refrigerator for more than six months. When it was rediscovered, there were no visible signs of mold in or on the pulp.)

(Top) On removing the paper from the bottom piece of cheesecloth, you will discover you have a sheet of handmade paper which bears the texture of the cheesecloth. If you find the texture of cheesecloth objectionable, now is the time to place the sheet between two pieces of woolen blanket, zinc plates, felts, pieces of canvas, gunny sacking, or whatever, and run the "sandwich" of handmade paper and its outer coverings through a clothes wringer or an etching press. Another alternative would be to place your "sandwich" in a bookbinder's press and give it all the pressure you can afford.

(Right) Forcing still more water from the sheet and imparting a felt or woolen texture by using a small bookbinder's press.

4.
WORKSHOP AND EQUIPMENT

*He hath not fed of the dainties
that are bred in a book;
he hath not eat paper, as it were;
he hath not drunk ink*

—Shakespeare, *Love's Labour's Lost,* Act IV, Scene 2, Line 25

VINEORANGE

The spaces and places in which the handmade papermakers I visited work vary from dark, dank, humid home basements to a particular room or rooms in their "castle" at ground level. There were also an owner-constructed mill (built with the gracious help of friends, neighbors, and apprentices); an enormous dairy barn in a constant state of reconstruction situated about a hundred yards from where the papermaker lives; and various industrial spaces in urban areas that required money with which to keep the landlord at bay.

The square footage also varied from numerous, miniscule, closet-like spaces to open, free, airy, high-ceilinged, factorylike areas with the kind and quality of space that is the very stuff of dreams.

All of the papermaker's mills visited possessed access to water and electricity; devices that could beat half-stuff, natural material, or rags; a vat or vats; moulds and deckles of various dimensions on which to form virgin sheets of paper; and a means of drying the several papers these hard working people manufactured.

None of the mills were alike in any way, though all contained similar devices to achieve their ends. (I cannot but remark on an article published in 1958 that noted that not one commercial handmade paper mill was in existence at that time in the United States!)

SPACE

For our purposes, since you are probably not now concerned with the problems of producing reams of handmade paper for sale to others, a garage, kitchen, or basement would be more than adequate—with the advice and consent of spouses, children, relatives, or others with whom you may live—unless they have different designs on the space you propose to turn into a one-person paper mill.

Remember that the space will be subjected to attacks of water in drips, drabs, drops, and occasional gushers in the course of making paper. Thus, a water supply is absolutely essential, as is electricity, for operating one or more of the pieces of equipment you may buy, purchase, or barter for something of value you no longer need. So, although it is not terribly neat to mention here, it would be most useful if you owned a number of mops, buckets, wringers, a sturdy pair of waterproof boots, and a rubber apron.

EQUIPMENT

Advice to future papermakers, with regard to equipment and tools needed, varies widely depending on whom you ask. But in general, don't turn down offers of anything that even remotely resembles a machine or device that may be of use now or later in your papermaking activities, especially if you are of the "pack rat" mentality.

BEATERS

Seek out surplus stores, junkyards, neighborhood garage or patio sales, used equipment and machinery companies, auctions, local antique shops, listen to radio "trades" or sales of used things, and watch similar programs on local or cable television stations. Consult advertisements, or, as a last resort, place an advertisement in your hometown newspaper requesting such items as: osterizers, blenders, meat grinders, food grinders or mixers, egg beaters, a mortar and pestle of fair size, a mechanical or electrical mortar and pestle, any other kind of mincing machine you can alter or make from parts of other devices. If you wish to own the Cadillac of beaters for your private use—and if you will not miss the funds—purchase a used or new one and one-half pound beater (which is a laboratory model of the 300-year-or-so-old Hollander—see Suppliers List).

Harold H. Heller, distinguished research consultant to the paper industry (and no relation to the author), noted the following on this subject: "If I wanted to go into the business of making handmade pa-per, I would build a hydropulper. These machines can be bought, ranging from 25 to 3 feet in diameter, but in essence they are big Osterizers or blenders. I think the hydropulper type of refiner would be the easiest to make in a home shop, but all kinds are big power users and would require more electricity input than would be found in the ordinary household supply circuits. There is nothing 'holy' about a beater—all the kinds of refining equipment do a little different kind of pulp fiber modification—but all can be brought to producing about the same results by controlling the through-put, . . . the amount of water used [that is] the fiber-to-water ratio, or [the] consistency. You can buy any type of pulp refining equipment in small sizes, but the cost is very high. I would try to get a used machine from one of the second-hand machinery vendors."

As in so many life situations, there is no substitute for experience; no book, no treatise will act as surrogate for the real thing; hopefully, after reading this work from cover to cover, you will not try to reinvent the wheel, but may confine your creative impulses to improving upon the particular tools, materials, and equipment used in making handmade paper.

THEORIES OF BEATING

To state it briefly: the first steps in beating rags or natural materials to form thin multiple sheets of consistent quality is to defiber the fibrous raw material, or to divorce the individual fibers by cutting or bruising (wetting) them, or both.

Fibers are made up of plant cells the lengths of which greatly exceed their transverse dimensions, not unlike spaghetti. Cotton fibers, for instance, average about 1 inch in length. Cotton linters, on the other hand, average about ¼ of an inch in length: they are more cylindrical, and their cell walls are thicker than cotton fibers.

When the cell walls of fibers are submitted to the scanning electron

(Right) Cotton linters, seen under a scanning electron microscope, magnified 130 times. (Below) Unrefined or unbeaten Kraft, softwood pulp, magnified by a scanning electron microscope 110 times. (Opposite Page Top) The same Kraft pulp—after 50 minutes of beating. Note the swelling of the fibers, the microfibrils which will assist in interlocking the sheet, the seemingly greater flexibility of the fibers, and a readily discernible increase in the interfiber bonded area. (Opposite Page Bottom) Cross sections of Aspen fibers (Populus tremuloides) magnified 2200 times. All courtesy R.A. Parham and H.-M. Kaustinen, The Institute of Paper Chemistry.

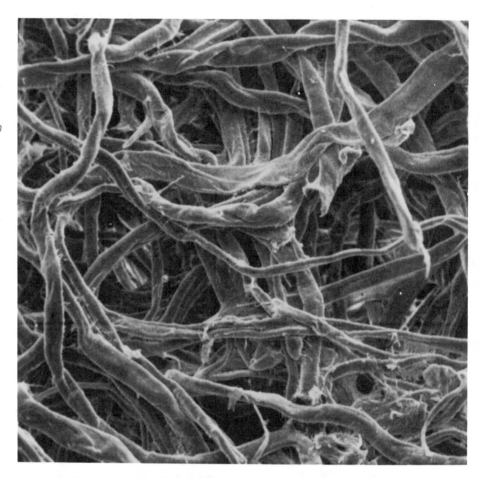

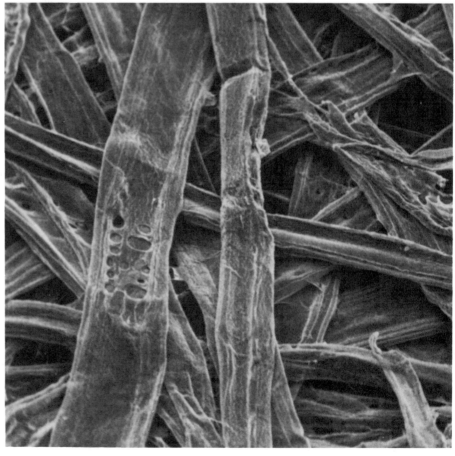

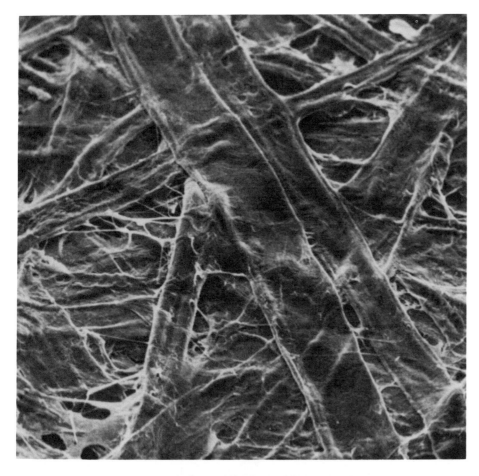

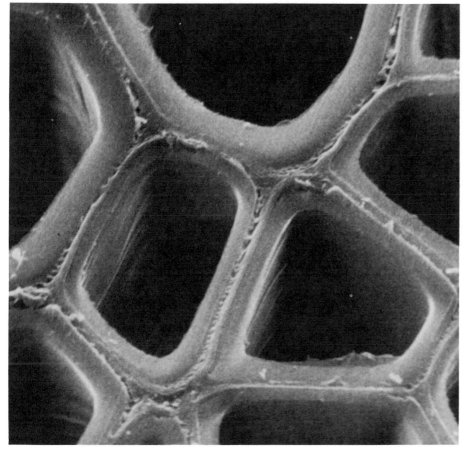

microscope and magnified 2200 times, the cross sections can clearly be seen (see the illustration of Aspen fibers). Differences in cell wall thickness and cell diameters—and length of the fiber, which is not visible in that photo—seen in cross section are due to earlywood (spring growth) or latewood (summer growth). Latewood tends to be thicker-walled, and its fibers are more rectangular than earlywood. There appears to be considerable disagreement among experts as to what occurs during the beating process.

In general, it is believed that beating will increase the mechanical strength of the end product or, to state it in a more "scientific" manner, beating will "increase the bonded area between fibers by making them more flexible so that they deform, preferably plastically, under surface tension forces during drying." (Emerton in Bolam, p. 41.)

John R. Peckham, Research Fellow at the Institute of Paper Chemistry, confirmed certain of the difficulties handmade papermakers face in his views on "Stock Preparation in the Art of Handmade Paper":

"No area of activity is more critical to the maker of paper handsheets than proper stock preparation. And probably none is more difficult to provide for in the matter of efficient mechanical aids. A small-scale beating engine or bar-type refiner is almost impossible to come by through the usual channels, and there are few viable alternatives to the infusion of energy for fiber fibrilation that a beater produces. This article will not make any contribution to the basic problem, but will attempt to address the reasons that some kind of a compromise refining solution must be found before the hand papermaking can put out a quality sheet.

"Even if the fiber furnish is adjusted to include a preponderance of wood fiber or well-prepared waste paper, the presence of even a little poorly refined rag or cotton linters stock will seriously flaw the paper's appearance. The reason for this goes directly to the natural cotton fiber produced in nature. Whether textile (staple) cotton or cotton linters, the fiber itself has a natural twist that tends to bring the fibrils into close contact on the surface of the cell. In addition, the process for making thread requires that the fibers be gathered together and twisted, which imparts a further hardening of the surface. The threads are woven to make fabric, which becomes rags, and the papermaker must face the problem of undoing all of the mechanical processes just described. Even then, he faces the task of so cutting and scoriating the cell surface as to free fibrils to give added bonding surface. In cotton linters pulp, the artifacts of spinning and weaving are eliminated, but, because the fiber is basically much shorter than staple cotton and possesses a much lower length to thickness ratio, the need for mechanical refining is no less critical.

"The need for beating or refining of paper pulp fibers is not limited to providing more bonding sites, but is also a critical variable in the preparation of a well-formed sheet. Too much fiber length encourages flocculation and the end result is a "wild" formation. Judicious use of the beater to brush out, hydrate, and shorten the fibers is the most responsive tool that a papermaker has. There is no substitute for experience and judgment in determining the beating endpoint, so the inexperienced are forced to experiment and reject as unsatisfactory much paper that shows the effects of too little or too much beating."

Papermakers, in the past, may not have known about the theories of fibers in the cell walls of plants and trees, how the outer walls are battered thus allowing the individual outer fibers to imbibe all the water they can accommodate. Similarly, they were probably not aware that as the fibers are repeatedly bent by the forceful shearing action that takes place in the small clear-

A Concise History of the Beater

*I*n the beginning, raw material for papermaking was soaked in water and then beaten between weighty stones. In time, a crude form of mortar and pestle made of wood or stone was substituted for the previous Neanderthallike approach. This was followed by a stamping mill, which was nothing more than a water, wind, foot, or horse-power-operated mortar and pestle. Then, when it was technologically feasible to provide a series or a gang of stampers, a true mill was developed.

Early in the seventeenth century, another technological improvement in defibrillating raw material was invented: it is suggested that this new invention, called a *kapperij* (a chopping mill) originated in the Zaan region of Holland. A *kapperij* is a slowly rotating vat containing numbers of spearlike hammers (shod with sharp iron points) that chop rags, sails, or other materials into useful pieces of rags by moving up and down through the manmade fabrics.

This chopping mill was followed by the forerunner of the hollander beater: the *kollergang*. And, finally, between 1660 and 1673, what we foreigners term, the Hollander (see pages 62–64) beater and the Dutch called a *bak* (trough) was developed in its present form and named, quite clearly, after the country that invented and refined it. Until the Hollander, only a coarse, brown paper was made— wrapping paper. (Much of this information derives from numerous articles by one of the world's foremost paper historians, Henk Voorn. See especially, H. Voorn, "Last Wind-Driven Paper Mill. . ." *The Paper Maker,* Vol. 31, No. 1, p. 3, 1962.)

Although the Hollander is still regarded by many papermakers as the ideal beating instrument, especially for rags, its "intermittent" mode of defibrillating rags brought about the invention of "continuous" refiners, among which is the cone-type refiner created by Joseph Jordan (d. 1903) in 1860, still held in high regard. In addition, between the years 1790 and 1885, there were no less than 97 separate patents issued in the United States for pulp beaters with bed plates, the most interesting of which were the Hoyt and Umpherston machines.

Earlier, the brothers Fourdrinier (Henry and Sealy) received a patent for a paper machine which still bears their name (see page 26) —although the technological "monster" we know today bears little relationship to their original device, save in concept—and is truly responsible for the paper revolution. (Imagine a Congress or a Parliament without paper trying to conduct its day-to-day-business; fancy a people, anywhere in the world, without its daily newspaper; picture a commercial society without paper to promote its goods and services or to disseminate information; conjure up a school system without paper with which to provide textbooks and tons upon tons of educational materials with which to educate its students. Impossible! Since we are here concerned with handmade paper, I will leave this subject for others and return to our immediate concerns.)

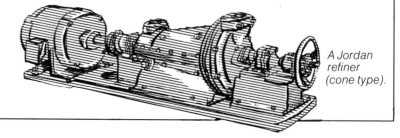

A Jordan refiner (cone type).

DER HOLLANDER

Being the beater of the Oak Park Press & Paper Mill at Wichita, Kansas, U.S.A.

1. Tub; 18 x 36 x 14 in.
2. Roll; 13 in. dia. x 8 in.
3. Bed-plate
4. Back-fall
5. Mid-feather
6. Motor; .5 hp.
7. Clearance adjustment
8. Drain
9. Roll hood (removable)

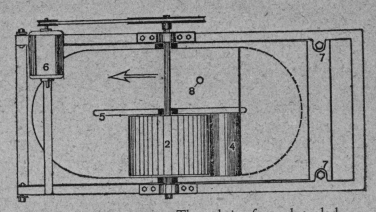

The tub is of wood, sealed with fiberglass, while the roll and bed-plate are of hard aluminum alloy. A steel chassis supports the motor and roll assembly, which is adjustable for roll-to-bedplate clearance. Roll turns at 290 r.p.m. Pulp flow rate: 6 circuits per minute. Tub capacity: 10 gallons.

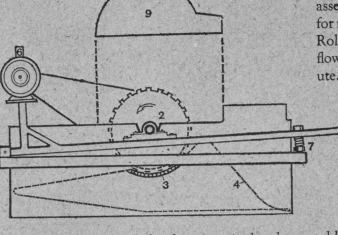

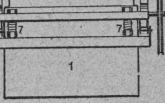

This diminutive broadside has been conceived and executed by Jim Yarnell as a keepsake to commemorate the 1st Papermakers' Conference, at Appleton, Wisconsin, November 21 - 23, 1975.

"Der Hollander." Received at the First International Papermaker's Conference at Appleton, Wisconsin, November 21–25, 1975.

ance between the roll and the bedplate and as they continue to imbibe water, the fibers are internally fibrillated. I strongly suggest that intuitively the papermaker sensed that the plasticized, fibrillated fibers interlocked during his four-way shake in forming a sheet of paper —he could not help but be aware of and sensitive to the considerable surface tension forces at work when he was in the process of making a sheet, if he remembered the days of his apprenticeship when he allowed the mould and deckle to become completely submerged in the vat and tried to pull it straight up.

To recapitulate, generous beating of cellulose fibers in water brings about a multiplication of individual fibrils. These fibrils cross over each other and intertwine as a result of the vatman's shake, the couching and drying procedure, to produce a sheet of paper; the strength of that sheet of paper depends, to a great extent, on the total number of bonds between the interfaced surfaces of the fibrils. Further, were you to examine just one of these fibrils under a very high-powered electron microscopy, you would discover that there are millions of microcrystals hooked together. And, pursuit of this idea will land you squarely in the midst of microcrystal polymer science, which, again, is not the proper concern of this book.

BUILDING A HOLLANDER

That piece of equipment called a Hollander, be it a Valley Beater, a Noble and Wood, or any one of a number of others, is made of a large oval trough or tub of cast iron, bronze, stainless steel, concrete, fiberglass, or wood, which may be unlined or lined with copper, tiles, lead, epoxy resin or other rust-free material. It is divided in half by a midfeather, around which the rags or natural material circulate. The beater roll, usually placed between the midfeather and one side of the tub, is a cylinder made of aluminum, wood, or other rust-free mate-

rial from which projects a number of metal bars or knives all around its perimeter. The stuff (untreated pulp) is processed by the rubbing or shearing action caused when the beater roll is rotated against a set of bedplates (knives) fastened to the floor of the tub or trough. The clearance between the roll and the bedplate is adjustable by bolts, which allow for changing that clearance during the beating process (see Chapter 2).

The bedplate bars, usually made of bronze or carbon steel, are placed to meet the beater roll at a slight angle to allow a shearing, or tearing, action rather than a cutting or chopping motion, i.e., either the bedplate knives are placed diagonally in the bedplate or the knives are bent at a slight angle.

The design of the Hollander is such as to provide a closely felted paper, because *the fibers are not all reduced to the same length*. This is caused by the fact that the stuff at the midfeather circulates more frequently than that at the outside.

"Der Hollander" on page 61 is a complete, visual presentation of the particular Hollander designed and constructed by James Yarnell. The information contained on that diminutive sheet would enable an enterprising individual, given time and the gift of craftsmanship, to construct a beater *after* having made paper for an appreciable time (after having developed an appreciation and respect for the myriad problems involved in making handmade paper)—if the person were a genius at seeking out and finding useful scrap from various surplus stores and junkyards.

USING THE HOLLANDER

Paper is made in the beater–Anonymous

The first user of the Hollander would have noted ". . .that he got a different sort of paper if the bars of his Hollander were old and blunt, compared with the paper obtained when the bars were new and

sharp. In the first case, the paper would tend to be hard, strong and parchmenty; in the latter case, it would be softer and more absorbent." (Cottrall in Bolam, p. 5.)

Also, in the first instance (with blunt, old bars) the pulp "drained much slower on his hand mould than the second type of stuff and moreover was slippery when squeezed—the water coming away from the mass only with difficulty. What would be more natural than to call this type of stuff 'wet stuff' or 'slimy stuff'? On the other hand, stuff obtained from a Hollander fitted with new sharp bars and stuff treated only long enough to complete the defibering process, were found to drain relatively quickly on the hand mould and to give up water relatively easily when squeezed—this was naturally called 'free stuff' in contradistinction to 'wet stuff.'

"Thus were born the terms *freeness* and *wetness* as descriptive terms of stuff quality." (Cottrall in Bolam, p. 6.)

As has been noted by John R. Peckham and others in last several hundred years, there is no proper answer to the question, "When is the stuff in the beater ready for papermaking?" other than experience.

The same raw or manmade material in different percentages of fiber to water, different speeds of circulation around the trough or tub (through alteration of the speed of the beater roll), different spatial relationships between the revolving roll and the stationary bedplate, different types or kinds of rolls and bedplates, and different lengths of beating time will produce very different end products.

To further complicate this "simple" process, the nature of the water employed in the beater and the beater instrument itself are still other variables to be taken into consideration. Finally, the bars of the beater roll may be sharpened at the edges and beveled down to about $1/16$ of an inch for beating free stuff (thick, soft bulky papers) or the bars may be blunt and about ⅜ of an inch for wet stuff (think, strong parchmentlike papers) or the roll could be changed and a substitute basalt lava stone roll inserted in its place to make a very wet and highly filbrillated stock for special purposes.

If you are fortunate enough to acquire your own beater, here are a few technical suggestions:

1. Find your "scratch mark" (the point at which the beater roll and the bedplate just barely touch each other), and make a visible mark on the adjustment bolts that raise or lower the beater bar.

2. Similarly mark the fully raised position, a "brushing" position (see below), and so on. All of these positions are used at various times in the making of particular sheets of paper. For instance, for very soft, almost blotterlike, or other bulky papers, you would want to furnish the beater very lightly and lower the beater roll right down to the scratch mark at the very beginning of the beating to cut the fibers and create a very free paper. For parchmenty papers, you want a wet pulp, one that is just bruised and shredded fiber.

3. Usually, if you are following no special formula, you would feed as much rag and/or manmade material as will circulate in the water already speeding around and around in the Hollander. If the beater begins to labor under the load of stuff you have added to the machine, lower the beater roll immediately to provide a good grip on the half-stuff. This will provide a combing action, will shorten the fiber, and will make the half-stuff move freely to the point at which you could add even more half-stuff, if you wished.

4. If you do not remember to lower the beater roll periodically, you may produce an overwet pulp that you may find useless.

5. To help clear the beaten pulp of knots, raise the roll to the brushing

John Babcock's Hydropulper

*T*ake several parts Yankee ingenuity, an ability to see things that others overlook or discard, an unsatiable curiosity, plus a talent that allows and encourages some people to use their hands with tools and materials to shape anything into something else and *Voilà*! You have John Babcock's new solution to an old problem.

"My beater is a Babcock/Maytag. It's a hydropulper that pulps up about 10 gallons into perfect sculpture pulp and also great embossing/etching paper. My cost was $50!"

In addition to producing wondrously unexplored works in paper (see the gallery), Babcock converted this used Maytag washing machine into an unusual hydropulper for cotton linters that furnishes him with "good pulp in five minutes, though I usually run it for about ten minutes." See page 98 for more information on this artist's use of his recycled device.

Portrait of the Babcock/Maytag hydropulper. Photo Richard Kluge.

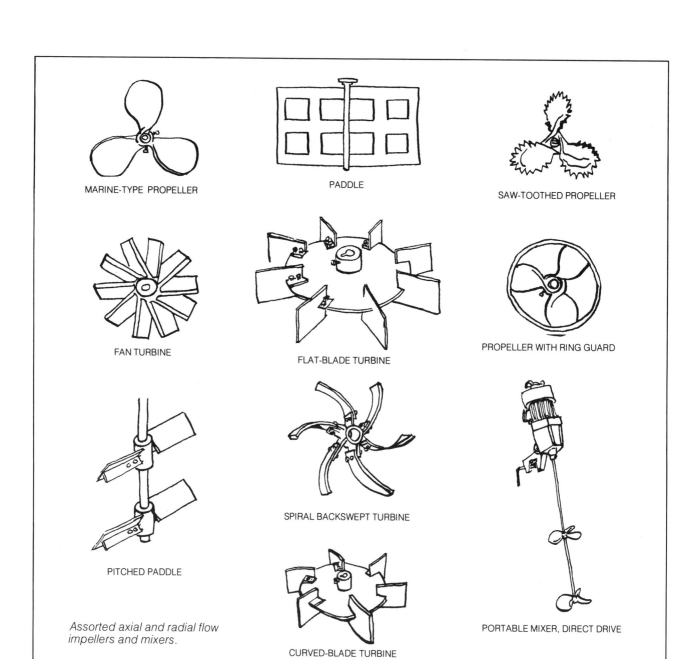

MARINE-TYPE PROPELLER

PADDLE

SAW-TOOTHED PROPELLER

FAN TURBINE

FLAT-BLADE TURBINE

PROPELLER WITH RING GUARD

PITCHED PADDLE

SPIRAL BACKSWEPT TURBINE

PORTABLE MIXER, DIRECT DRIVE

CURVED-BLADE TURBINE

Assorted axial and radial flow impellers and mixers.

Flow Patterns

Left to right:unbaffled, off-center;
unbaffled, angled, top entry;
unbaffled, off-center;
baffled turbine, on-center;
baffled propeller, on-center.

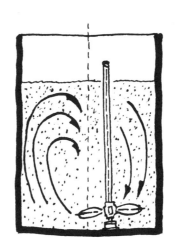

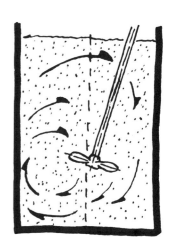

position—this position has no weight on the roll—for about 15 to 20 minutes.

Everything I have written here is a dangerous generalization. There are too many variables at work. If you happen to own a beater, see the bibliography for references to more detailed works. Let us move on to other areas of the papermaking process that are common to all makers of handmade paper, no matter what your equipment. But, we must include a unique do-it-yourselfer, first.

NOTES FOR THE DO-IT-YOURSELF HYDROPULPER BUILDER

The following information is contained in rather outmoded chemical engineering handbooks, not because the information is faulty, but because chemical engineers, by and large, are more concerned with major theoretical and practical problems today than with impellers or jet mixers for pulping linters on the scale that concerns us.

One of the first problems encountered, when Garner Tullis and I first tried our homemade hydropulper in Santa Cruz in a 30-gallon circular plastic garbage pail using a radial flow impeller (blades which run parallel to the axis of the drive shaft) was an intense swirl, the formation of a vortex, severe air entrainment, and the drenching of me with pulp from head to foot.

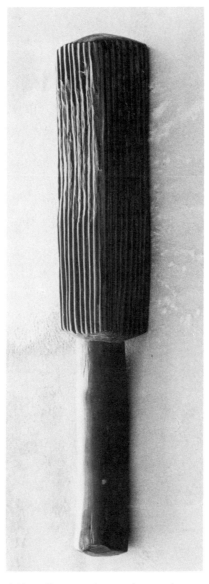

A Hawaiian tapa beater (second stage). Collection the author. Photo Merrylee Stephenson.

Reason? It appears: (1) we were using a tremendous amount of air pressure to drive our impeller; and (2) the plastic garbage pail in which we were trying to make pulp was not baffled (with vertically mounted 2 × 4s) to stop the swirl. A temporary solution was effected by squeezing the vessel into an oval shape and by driving the impeller counterclockwise.

Cure? Even two flat, vertical baffles, for our purposes, would have been sufficient, had they been opposite each other and each of them $1/12$ the diameter of the tank. They could have been made of well-seasoned and waterproofed wood, glued (waterproof glue, of course), and securely fastened to the inside of the container.

With a very large tank, four baffles would stop swirl admirably. Other solutions may be obtained with unbaffled tanks by mounting the impellers off-center.

Still other possible solutions are clamping a portable mixer (obtainable in sizes up to three horse-power) to the side of the mixing vessel or varying of any of the impellers or turbines shown here that appeal to your esthetic sense.

OTHER PROCESSES FOR MANUFACTURING PULP

1. You could, if you possessed sufficient energy, use a mortar and pestle to pound away at cotton

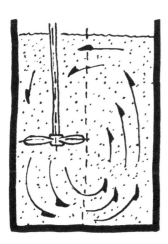

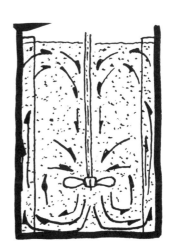

rags or plant materials, adding water from time to time, until you refined the material to a pulpy mass.

2. The photo on page 65 shows a beater (*I'e kuku*), a second stage beater possibly from Hawaii found in an antiques shop in Auckland, New Zealand. It is made of ironwood (or of some very dense, heavy wood), is 15⁵/₁₆ inches long; the block is 8¾ inches; each beating surface is 2 inches wide; and it has between 9 and 12 grooves on a beating face. It was used by Hawaiian women to make paper clothing and sundry tapa articles for their culture. You could, if you were so inclined, beat fibrous material on a flat log with this ''weapon'' until you obtained a pulpy mass.

3. If your woodworking skills were of a certain level, you could reproduce a scale model or a version of the historic stamping machine, and, thereby, increase your pulp-making capacities. But, since quantity is not necessarily the goal, let us review still other modes for macerating materials.

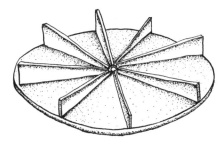

Tullis' experimental hydropulper blade.

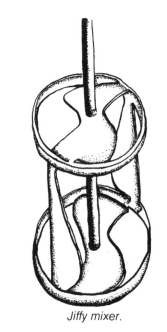

Jiffy mixer.

4. An eggbeater would probably be a useful tool with which to pulp fibers, if that material were facial tissue (not wet strength), well-soaked photographic blotters, or filter paper (which is, practically, pure cellulose).

5. My daughter, Jill, used our household food blender to create pulp from one inch, well-soaked squares of old cotton sheets. With a great deal of water and just a small amount of cotton squares (plus a tremendous amount of patience), she was able to create sufficient pulp with which to make a sheet of paper about 10 × 12 inches in size on an inexpensive expanding screen sold by hardware stores.

6. Garner Tullis provided me with my first hydropulper blade with which to experiment. If you use something like this, you must purchase linters or half-stock for hydropulping. You may wish to macerate the half-stock or linters in a plastic garbage pail with a ½ inch electric drill of variable speed and direction.

A manpowered stamping machine. Jacques Besson, Theatre des Instrumens Mathematiques & Mechaniques. Lyon, 1579, pl. 25. Detail.

Excerpts from Joseph Wilfer's History of Moulds

*I*n conjunction with an exhibition of 2,000 years of mould-making held in 1973 in Madison, Wisconsin, Joseph Wilfer did a vast amount of research on the history of the tool without which there would be no handmade paper.

Many individuals have speculated upon the materials and use of Ts'ai Lun's first mould. "It is not known if these first moulds were dipped into a vat or if the fibrous slurry was poured directly onto their surfaces. . . ." Suffice it to say that these early papers had to be air-dried, and one mould, therefore, was required for each sheet of paper.

Technology marches on. "In the eighth century the Persians devised a mould from which a wet sheet could be removed and dried, allowing the papermaker to make an unlimited number of sheets from a single mould." Wilfer compares these Persian moulds to bamboolike place mats held in a rigid carrier when forming the sheet. It is easy to visualize the removable "mat" made of bamboo strips placed side by side creating laid lines and horsehair used to tie the bamboo strips together forming chain lines, perpendicular to the former. These materials, by their very nature caused differences to be formed in the thickness and density of the paper. Thus, truly they were the first watermarks, as could be verified by holding the sheet up to the light of the sun.

When the western world became the recipient of the mould, it arrived as a rigid hardwood frame supported by a series of ribs across which was fastened a layer of "uniformly shaped and sanded fibers."

Wilfer goes on to recount the changes in the mould covering from natural fibers to iron wire, bronze, copper, and brass, coincident with the birth of wire, wire-drawers, and wire-smiths.

"Not until the eighteenth century when papermaking had spread almost over the entire world did the next significant development take place. Historians differ but many believe it was the famous printer, John Baskerville (1706–75), who, in search for a more perfect sheet on which to print, thought to raise the surface wires off the ribs. In doing so the draining was improved and the sheet that was formed was a good deal more uniform in density."

Wilfer believes that Baskerville was probably responsible for developing the wove mould. "This is simply a piece of fine woven wire screening that is placed over a series of backing wires."

His closing sentence is a masterpiece of understatement: "The same woven wire screen that made possible the fine handmades, in 1878 opened the door to the continuous web paper machine."

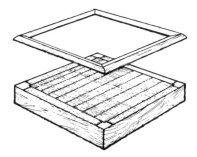

7. Using the same ½ inch electric drill, you may find a clay mixer to be a useful substitute in refining linters or half-stock. This particular device is called a Jiffy Mixer (see p. 65 and Suppliers List).

8. Others may find a high-speed, ¼ or ⅜ inch electric drill armed with a paint stirrer better suited to their requirements for beating linters.

9. You may wish to by-pass the entire operation and purchase already prepared pulp from one of many suppliers, most of whom require a minimum stated order. Perhaps, if the requirement is too high, you may interest a number of friends (see Suppliers List). The pulp is usually sent wet, wrapped in heavy plastic.

10. Some of the new kitchen gadgetry—those that blend, grind, beat, etc.—may also be useful substitutes if the proud possessor of said expensive machine will allow you to experiment.

11. You may wish to emulate John Babcock and build a hydropulper from a used washing machine.

12. You may not wish to emulate anyone who possesses so many useful skills and may prefer to buy a hydropulper (see Suppliers List).

13. You may purchase the "innards" of a Hollander-type beater (a beater roll and bedplate) for about one-tenth the cost of a new one and enjoy the privilege of making your own Hollander. (See Suppliers List and James Yarnell's beater in Chapter 2.)

14. If there is still one available by the time this manuscript becomes a book, you may be able to purchase a used Hollander-type beater by writing to various used-equipment dealers (see Suppliers List).

15. You can always order a new Hollander-type machine and wait impatiently for it to be delivered. Possibly this paper revolution will entice more individuals into the manufacturing of Hollanders to

bring down prices or, at the very least, to make small-scale models for individuals (see Suppliers List).

MOULDS AND DECKLES

The average size of moulds during the fifteenth century was 14 × 19 inches; the largest sheets measured 18½ × 26½ inches. There is no doubt whatsoever that a well-crafted mould and deckle is "a thing of beauty." Given the proper hand tools, patience, skill in the high art of carpentry and the working of metal, knowledge, and a good set of plans, there is no earthly reason why we all could not produce our own mould and deckle.

Realistically, however, not all persons wish to devote themselves to such projects; fewer and fewer want to spend their precious time in this manner; still fewer own or know how to use hand tools in the way they were intended by their makers. Further, the sometime papermaker, the occasional weekend papermaker, the once-a-year papermaker, and the person who wants to make it all day every day have, obviously, different requirements—even though all should be satisfied with nothing less than the best mould and deckle they can fabricate.

Here are a number of approaches to the making of moulds and deckles, any or all of which will work for you. Just select one or more of these procedures that, after reading, you know you can handle; that is one way of ensuring you will produce an excellent hand mould.

Don Farnsworth's easily crafted Honduras mahogany wove mould. To save yourself between $75 to much more than $500 per hand mould, depending on the size you wish to make rather than purchase, these are the simple steps to follow in making a reasonably long-lived wove mould courtesy of papermaker Don Farnsworth:

1. Cut lengths of well-seasoned hardwood (½ × 2½ inch) stock to create an open box. Veteran papermakers will recommend no wood other than Honduras mahogany—I suggest you use your own very good judgment in this regard. The *inside* dimensions of the box will, more or less, dictate the size of the finished sheets you wish to produce: 6 × 9 inches, 9 × 12, 12 × 18, or whatever.

2. Join the corners by dovetailing the ends into each other, and then glue (waterproof glue) and countersink the joints with flathead brass screws.

3. Obtain and cut to proper size (your *inside measurement*) a piece of plastic, egg-crate fluorescent light diffusion screening. (See Suppliers List.)

4. Strengthen the corners of the frame with brass or bronze corner angles fastened with flathead brass screws.

5. Using ingenuity, glue, or screws, tack or otherwise affix the fluorescent screening to the inside of your frame (rabbit the frame and rest the screening thereon?), *making certain that the top of the surface of the plastic egg-crate is level with the top of the frame.*

6. Fasten 8 mesh stainless steel screen to the diffusion screen. Tie it to the fluorescent light diffuser with 2 pound test fishing line, fine wire, or their equivalent—every 6 inches or so.

7. Using about 30 mesh brass, bronze, or copper wire screen, fasten and stretch it across the stainless steel screen, tacking it with copper or brass brads as you proceed, in the same manner as you would stretch a canvas across wooden stretcher bars. No wrinkles, please. Sew this down to the fluorescent light diffuser also.

8. To "neaten it up" (as my former Pennsylvania Dutch colleagues would advise), fasten a ½ inch brass strip around the entire frame and secure it with copper or brass brads. This brass strip will also prevent the raw edges of the screens from "grabbing" the felt and causing problems when couching.

9. Rub several coats of linseed oil *into the wood frame only*.

The Honduras mahogany deckle à la Henry Morris. Now that you have a mould, you need a deckle:

1. Using stock (Honduras mahogany) about 1 × 1½ inches, cut a ¾ inch wide channel about ½ inch deep in the underside of the hardwood about ¼ inch from one edge.

2. Cut the ¼ inch lip down to about ⅛ inch above the channel bed, allowing a slight step or foot which will bring pressure upon the wire when the deckle is in place. This will cut the deckle edge of your sheet cleanly.

3. Round the outside of the stock, and cut it into the four lengths necessary to make the deckle. *Cut each piece 1 inch longer than the side of the frame it is intended to cover.*

4. Note the unusual deckle joint construction in the illustration at right. Now, cut and try the joints before going any further.

5. Taper the long sides (front and back) of the deckle toward the mould face, leaving a flat edge on the inside surface. Depending upon the thickness of the paper you intend to make, the inside edge can vary from $3/16$ inch for book-weight papers to ⅜ inch for heavier stock.

6. Shape the sides of the deckle to fit comfortably and firmly in your hand.

7. Using a waterproof glue reinforced with thin brass angles secured by flathead brass screws, assemble the deckle as shown.

8. Test and fit. Test and refit, until the deckle, after filing and sanding, reacts as though it truly "belongs" to the mould for which it was fabricated. Finish with several coats of linseed oil.

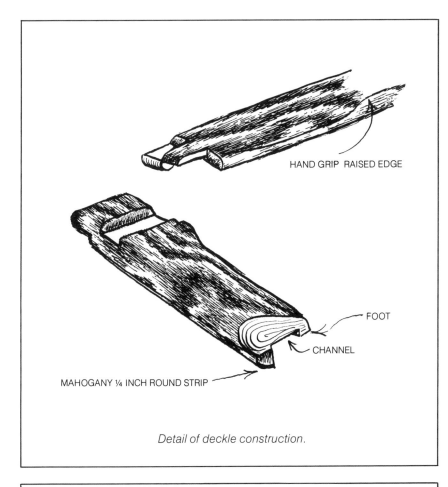

Detail of deckle construction.

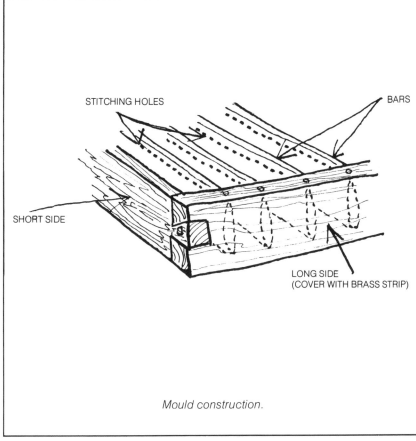

Mould construction.

9. Now, go back to your mould and make $1/16$ inch holes in the wire cover every $3/8$ inch along the four outer edges of the mould. They should be situated so they are covered by the deckle when it is placed on the mould. These holes will speed drainage when you are in the process of forming sheets.

10. Finally, attach a mahogany, quarter-round strip to the underside of the front piece of the deckle. This device is intended to keep excessive pulp from running under the deckle.

Inexpensive mould and deckle.

1. Select any kind of scrap wood (1 × 2 inch), cut to desired size to create an open box—up to 9 × 12 inches.

2. Butt the ends together (these are also termed L joints) by dovetail nailing (drive the nails into the corners at angles—slanted driven nails provide a better grip and a sort of hook effect).

3. Strengthen the corners with flat corner plates of brass or bronze, and fasten with flathead brass screws.

4. Waterproof the frame with a good waterproof glue.

5. Stretch a very fine mesh synthetic woven fiber across the top of the mould, as you would stretch canvas.

6. Make a simple wooden deckle from scrap picture frame material.

"Neighborhood frame store" mould and deckle.

1. Purchase four stretcher bars, which when glued together will give you an inside dimension, more or less the size of your proposed finished sheets of handmade paper.

2. Use epoxy or an equivalent glue and the wooden wedges that accompany the stretcher bars to join and waterproof the stretcher bars. Waterproof the *whole mould*. Be certain the stretcher bars are

square. If necessary, reinforce as described above for the other moulds.

3. Stretch a fine mesh synthetic nylon or polyester plastic screen across the top of the mould using brass brads or tacks to tighten the fabric. Avoid wrinkles.

4. If you have selected a standard size set of stretcher bars, there should be no difficulty in finding a standard, simple frame, which will act as your deckle and will fit properly over your stretcher bar mould.

Henry Morris' laid mould for all time.

1. Mr. Morris strongly suggests you use well-seasoned Honduras mahogany for your mould of 2½ × ½ inch stock.

2. Assume the inside measurements of the mould will be 11 × 17 inches.

3. Cut and miter the four pieces of wood accordingly (see page 69).

4. Mortise holes, at about 1 inch intervals, into the inside surfaces of the long sides of the frame (to receive the bars or ribs which support the mould face). The mortise holes may be shaped to fit the ends of the complete bars (an end-view would be not unlike 'an airplane wing) or, if you wish, shape the ends of the bars into dowel pins to allow the mortise holes to be drilled.

5. Dovetail the corner joints and secure them with countersunk flat-head brass screws, as shown.

6. Fashion the ribs or bars from clear pine stock ⅛ to ¼ inch thick. Taper (as in that same airplane wing) to about ¹⁄₁₆ inch at the top. The bars or ribs should fit the mortises snugly. When assembled, finally, use Weldwood or any other phenolic glue for certain waterproofing.

7. "In mounting the bars, it is important that the top (the ¹⁄₁₆ inch ends) surfaces all form an even plane with the top of the frame, since the mould face must be evenly supported to withstand the stresses of vat use. Very fine stitching holes should be drilled or pierced about ¼ inch below the top edge of the bars at intervals of about ⅜ inch." (Henry Morris, *Omnibus,* p. 11.)

8. Affix bronze or brass corner angles to the corners to strengthen the frame and protect it from wear. Using brass brads attach ½ inch half-round strips to the undersides of the frame for further protection.

9. "While laid moulds built for commercial use often use two layers of wire covering, a much simpler facing will suffice for the hobbyist. The easiest way to obtain a laid facing for your mould, of course, is to purchase a piece of used laid dandy roll facing from a paper mill. In my experience, this is strong enough to be used by itself as a mould covering. It can be attached by sewing it to the bars with fine wire (about .005 inch or so), with the edges secured under thin brass strips nailed around the top of the frame, just as a wove cover is attached." (Henry Morris, *Omnibus,* p. 15.)

THE VAT

Jost Amman's woodcut, the earliest known print with regard to papermaking, reveals a vatman, vat, a Sampson or press in the upper right, the stampers of the mill in the upper left, and the papermaker's devil? rushing to the left.

We can gain some useful information from the print with regard to the vat: it should be at waist level; it can be made of wood and be round or oval, it is much larger than the mould and deckle; and it is larger in circumference at the top than at the bottom.

The tub or vat has, at times, been constructed of stone, copper, iron (though lined with lead to prevent rust), plastic, cement, or a marine plywood rectangular object of various dimensions—it varied in size from that of a bathtub to about 5 × 7 feet, to the enormous sizes required by today's commercial machinery.

Vats for small handmade operations, though, were usually in the 5 × 7 foot range. The front and back walls of vats for handmade paper normally slanted inwards (to prevent the vatman from back strain when he was lifting large moulds and deckles full of pulp). In early times, a hog at the bottom of the vat agitated the pulp to keep it (the fibers) from settling. Today, papermakers merely roll up a sleeve, plunge their hand and arm into the creamy mass to stir things up, and remove their arms with the telltale traces of their calling visible and quite fixed to their hairs. Also, there was a wooden platform (a bridge) across the back of the vat perpendicular to which was accommodated a stay. Above the stay there was an asp, horn, or ass. Of these

Henry Morris' well-used vat. Photo the author.

The Papermaker *by Jost Amman, 1568.*

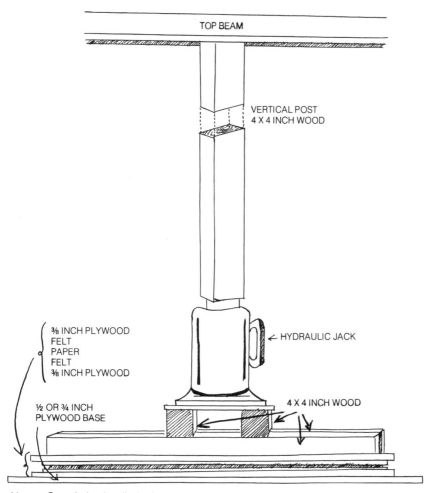

TOP BEAM

VERTICAL POST
4 X 4 INCH WOOD

⅜ INCH PLYWOOD
FELT
PAPER
FELT
⅜ INCH PLYWOOD

← HYDRAULIC JACK

½ OR ¾ INCH
PLYWOOD BASE

4 X 4 INCH WOOD

Nancy Genn's hydraulic jack paper press.

Henry Morris' press and felts. Photo the author.

latter three terms, the user employed the name in common usage in his culture.

I have seen homemade vats of marine plywood bathed in waterproof cement that worked efficiently, as well as inefficient ones of various materials. What is required, however, is any tub or vat, homemade or otherwise, that will be larger than the largest mould and deckle you intend to use. It must be waterproof and easily cleaned, and must be able to be set on wheels if you do not have a great amount of space and need to move things from one place to another. Again, the choice is yours.

DRYING THE HANDMADE SHEET

The following discussion is by Harold H. Heller: "In drying the hand sheets it is not necessary to be historical. In olden times, the wet sheets were taken from the pile of couched paper and stacked up in an even pile and pressed with heavy weights or with a screw press. All the water removed had to flow from the center of the pile toward the sides, or edges of the sheets. The edges of the sheets were wet at first, and, if allowed to stand for days, the edges were dry and the center of the sheets were wet. If these sheets were later hung up on a line to air dry, they would curl and cockle but not lie flat. Later, the damp sheets were piled up with a piece of wool felt between each sheet and again pressed. This would also give a more uniform water content to the sheet and would also press a surface pattern into the sheet depending on the woven pattern in the felt. (The term felt is used for a woven fabric, but originally the material was a wool felted material with no pattern.)

"A much more uniform initial dryness can be obtained in the paper by running it through a set of wringer rolls or press rolls while carrying it on a felt. The quantity of water removed from the sheet depends only on how completely the interstices between the pulp fibers

Commercial mills store and sell pulp, half-stuff, and linters in many forms. Pulp may also be purchased fully beaten and wet. The cube, here, is pH neutral wood pulp; the "sheets" are pH neutral cotton linters, a by-product of cotton ginning. Both require further beating before use in papermaking.

are closed up. It often takes pressures like 500 pounds per square inch to get a paper dryness of 65 to 75% dry content. After removing the water to this degree, the sheets can be re-pressed to give the desired surface pattern or texture. For a watercolor painter, this texture is all important. To the papermaker, the flatness of the dry sheet is very important. It is my belief that the modern handsheet maker could get these desired properties with a rolling press and subsequent texturizing process, with less trouble than with the old pressing in a big pile.

"Most paper machine felts are now made of synthetics and contain no wool. This is because the nylon (for example) has much more wear resistance than natural wool. It may be difficult for the handmade papermaker to get wool felts for texture controls of the sheet surface, except by having the felts made to order by one of the big pa-

permachine felt makers."

So, one senior person in the paper field, for reasons just quoted, suggests you use a clothes wringer and an etching press and that you dry your paper between wool felts.

If the texture secret is wool, would there be old blankets of that material at hand? But, suppose you wanted a texture similar to that of canvas for your paper? You could, at the last pressing, press your sheet between linen or cotton canvas. It would be simple to discover the various textures you could get with clean gunny sacks (the kind numbers of us used to "keep" our catch when we went fishing on the Pacific coast); and it would be worth exploring the world of all contemporary textile fibers for other effects.

If you are fortunate enough to live in a sunny area, you may also decide, one time, to see what effect comes without benefit of any pressing device: dry your wet

sheet right on the mould, out-of-doors, in the sunshine.

Some of you may decide to squeeze dry a post of paper in a bookbinder's press or a standing press, placing 5-ply marine plywood sheets under and over your felted sandwich of wet papers. Still others may wish to find a rebuilt hydraulic builder's jack, or a heavy-duty scissors jack with which to obtain a strong force to drive out water from freshly made sheets. Given a very strong framework (say that of a very tall standing press) along with 4 × 4 inch and 4 × 6 inch stock plus a great many sheets of 5-ply plywood, it would be a simple matter to jack up or jack down the post against an immovable (hopefully) crossbeam or bar.

Still further, it may be useful to attend used machinery auctions, including army surplus or what-have-you, searching out ram presses, moulding presses, and all other devices that appear to you to do what it is you want done.

Obviously, if you are making very small sheets as a hobby, you can always press and dry your paper between blotters with an electric household iron. When doing this, *remove yourself from the wet area* to a very *dry* place. (We do not want to encourage or be responsible for electric shocks, especially if the electric cord is frayed or worn in spots.) Allow the iron to reach warm to medium heat, and press down upon the blotters. Don't slide the iron across the blotters; they may get torn up rather quickly. Drying may be speeded up turning the blotters and ironing first one side and then the other. Repeat until fully dry.

NATURAL OR MANMADE MATERIALS AND SUPPLIES
Improvisation in papermaking, as in so many areas of the arts today, appears to be the rule rather than the exception. Make substitutions for these suggested items whenever your desire and pocketbook do not balance out, when you want to know, "What would happen—if?"

Over two centuries ago, Dr. Jacob Christian Schaffer (1718–1790), in a search to discover new materials to substitute for linen and cotton rags, published a rare six-volume treatise, with specimens contained in these books of the following: poplar down, wasp's nests, sawdust, pine chips, beechwood, willow wood with and without some rags, tree moss, poplar, hop tendrils and wood, the inner and outer bark and wood of grape vines, and combinations of all of those materials. These samples appeared in Volume I, which was published on January 15, 1765.

Within a space of two and one-half months, this eighteenth-century experimenter published another volume in which he used aloe leaves, barley, hemp, mulberry and its inner bark, stinging nettle, clematis wood, willow, bull rush, earth moss, the wintered leaves of trees, cabbage stalks, scraps of previously made colored and uncolored specimens, and roof paper derived from tow.

On November 3, 1765, Schaffer made paper and published his third volume of specimens derived from: seed down, burdock stalks, thistle seed down, Cyprian asbestos, thistle stalks, lily-of-the-valley leaves, water moss, Bavarian turf, Hanoverian turf, and a variety of thread that was knitted, woven, colored, plain, and so on, made from poplar down.

To celebrate New Year's Day in 1766, Volume IV appeared which contained paper specimens from Artemisia wood, Indian corn, young grape vines, silk plant, mallow, orache, spruce fir wood, lace made from aloe fiber, and a specimen made from scraps of all the papers in that volume.

About three and one-half months later, still another volume appeared with paper specimens from genista, pine cones, potatoes (the peels and the vegetable itself), shingles, and scraps of all of the papers in that volume.

Some four years later, Dr. Schaffer produced paper from reed, horse chestnut leaves, Brazil wood, tulip leaves, linden leaves, walnut leaves, genista, yellow wood, and a sheet made from scraps of all of the above.

MISCELLANEOUS SUPPLIES AND MATERIALS
Felts of wool, woolen blankets, canvas, burlap, blotters, roving, and zinc plates are required for couching, pressing, providing various textures to the sheets when they are still damp, and calendering for the equivalent of cold or hot-pressed surfaces.

Large rectangular sponges to be used for felting.

With regard to dyes, turn to the discussion on pages 79–80 and then to the Suppliers List.

Watermarking wire. (.030 inch soft brass wire).

Sewing or tying wire for mould screens (.005 inch).

Copper or bronze wire wove cloth (about 30 mesh).

Laid wire cover for mould (from makers of dandy rolls). See Suppliers List.

Brass brads, nails, angle corners, and screws.

Assorted woodworking and soldering tools.

Lint collected from your dryer and everyone else's.

Various chemicals, loaders, fillers, etc.—including titanium oxide, EPK kaolin, Aquapel, and Methyl Cellulose—for brightening, strengthening, and/or preventing shrinkage of paper.

Large plastic garbage pails with tight-fitting lids.

Mops.

And, for yourself, rubber boots and a rubber apron.

ON WATERMARKS
Every freshly moulded sheet
like wafers bears the print of mesh
and leaves the maker's crest ingrained.
One may choose a coiling snake,
perhaps a bunch of Bacchic grapes,
a rose, a bell, a cockerel,
or probably a monogram.
One could tabulate a list
with caprices of every man's fancy
but that is not the point of this.
However, for the purchaser,
the watermark proves quality
symbolic of the craftsmanship.

These lines were penned by Father Jean Imberdis, S.J., a native of Ambert in Auvergne, in 1693, and they still fascinate papermakers, historians, and others for various reasons. The historians are engaged in attempting to trace the origin and date of manufacture of particular sheets; others are engaged in searching out hidden symbolic meanings behind the watermarks —a few suggest watermarks are merely symbols with which to identify mould sizes and, thus, the paper made on them. Still others eye them with high regard for their demonstration of craftsmanship as a trademark; some state categorically, they were mere "fancies" of the papermakers. Finally, there are those whose concerns with watermaking relate solely to security.

It must be pointed out that watermarks are not really caused by water. If a papermaker, through carelessness, allows a drop of water to fall, inadvertently, upon a freshly forming sheet of paper, he has, indeed, created a kind of watermark. Some people would call that mark a "papermaker's tear," would be most unhappy about this unfortunate accident, and would discard the sheet. (On the other hand, there are certain artists using paper as an expressive medium who actually woo such "accidents" deliberately for esthetic reasons.)

But watermarks, to keep using this inappropriate term instead of the more reasonable French word *filigrane,* are made to appear in a sheet of paper as translucent figures, emblems, or letters as a result of raised wire (or moulded) designs affixed to the moulds.

Thus, we note that the watermark, which sits *on top of the mould* will, by its very nature, vary the thickness of the paper in the *forming* of the sheet. The pulp on the mould, fenced in by the deckle, will be thinner or thicker according to the design of wires. When held to the light, after forming, lighter or darker areas will be noted.

Analysis of numbers of watermarks lead to the conclusion that designs for watermarks should not utilize the too-acute angle, small circles, overly long straight lines, and solid areas. It is assumed that the makers of watermarks knew what they were about and that they did not try to go beyond the limitations of the process of papermaking. It is suggested that incorporation of any of the above-mentioned taboos in a watermark was looked upon as cause for undesired variations or problems in the forming of sheets.

Dard Hunter divided watermarks into four categories:

1. The earliest known ones, which include: circles, crosses, knots, three-hill symbols, triangles, and other devices that could be easily twisted from wire. (These were in use until the first quarter of the fifteenth century.)

2. Literally, thousands of designs based upon man and his works.

3. Flowers, fruit, grain, plants, leaves, trees, and vegetables.

4. Domesticated, legendary, and wild animals, as well as crabs, fish, scorpions, snails, snakes, turtles, and all sorts of insects.

Whatever the ultimate meaning of the watermark (and there are those who subscribe to all sorts of interesting theories about secret societies using this means of communication to express heretical views, among the more fanciful ideas), there appear to have been none in use in the Orient—perhaps because thin, rigid bamboo strips used in the moulds in the Orient do not appear to lend themselves to free-flowing imagery. Even though the rigid mould and deckle developed in the western world would have, early on, supported the use of watermarks, they do not appear until 1282 in Italy and became generally employed during the fifteenth century.

As to the use of watermarks for "playing detective" to determine the country, date, mill, and perhaps the master vatman who made the sheet, it is disheartening to learn that even the great William Caxton (1422?–1491), the first printer in England, never used papers of one watermark design in his books. To find paper of consistent quality and thickness, he would use from 15 to 20 differently watermarked papers of various dates and presumably from different mills in Holland. (One exception to the above is a rare and perhaps unique sheet owned by the Paper Makers' Association of Great Britain, which contains Caxton's printer's device and John Tate's watermark in the same sheet!)

It would have been common in the past, as it is today, for paper-

<hr>

False Watermarks (A Note)

*T*he nineteenth century was replete with individuals who produced numbers of inventions of interest to papermakers. In France, for example, Messrs. Davanne, Jeanrenaud, and Maquet independently arrived at a method of producing a "water-mark" made of transparent or translucent lines in the sheet by running a hardened gelatine relief through a press (similar to an etching press) under great pressure, in conjunction with a blank sheet of paper. The result, on the blank sheet, was the mirror image of the design on the carbon print —a "water-mark."

It appears that these gentlemen, including Messrs. Marion, Gobert, and still others, "stumbled" on to a process patented, prior to 1870, by Walter B. Woodbury (1834–1885) of England.

Mr. Woodbury's invention produced a perfect, transparent photographic watermark: "By reflected light the depressed portions appear of a deeper tint than the rest of the paper; if a transparent positive instead of a negative be used in the first place, a pretty, delicate grey positive is produced. . . . [T]he only wonder is that as yet such a good idea remains to be commercially utilized." (I am indebted to Professor Bill Jay of Arizona State University for this note on "Watermarking Paper by Photography." *The Photographic News*, Vol XIV, No. 608, April 29, 1870.)

makers to purchase old moulds and deckles with watermarks on them from wherever they could be obtained—on the demise of a given mill, as a souvenir, because they were bargains, or just because. Thus, it is not surprising to find that numbers of old English and other European moulds were used in North America (and are still in use by some papermakers). Pity the poor student of the history of papermaking discovering, say, a watermarked paper from a Brewer's mould (an English firm that has been making moulds and deckles since the eighteenth century) in someone's handmade paper mill in Apache Junction, Arizona, in the 1970s! Or, better yet, a laid mould from the hands of one of our foremost contemporary American papermakers and publishers, Henry Morris, which produced a sheet of paper in a paper mill in Pondicherry, India!

Although it seems far-fetched at first, there appears to be a relationship between the so-called art of war and the art of papermaking. The "secret" of papermaking was supposed to have been wrung from Chinese prisoners of war by their Arab captors in the middle of the eighth century. About eight hundred years later, Austrian soldiers invaded Antwerp, a well-known paper center, and forced its paper merchants to flee to Holland. Holland was invaded by Austria and Spain and turned for assistance to Queen Elizabeth. She responded by sending one of her armies led by the Earl of Leicester. In the latter part of the seventeenth century, the armies of Louis XIV invaded Guelderland. The paper mills were dismantled and set up in the Zaan district, near Amsterdam, for safety.

A hundred or so years later, a state of undeclared war existed between counterfeiters and the Bank of England with regard to the banknote paper employed by the government, watermarks, engraving, and printing. The counterfeiters were, seemingly, undeterred when

members of their talented group were convicted and executed; in fact, between 1797 and 1817 there were 870 prosecutions in England for banknote forgery for which 300 persons of artistic talent were put to death after conviction. During the next year, 1818, another 32 forgers were executed. The Bank of England, in an effort to defend and protect its currency, was forced to employ no less than 70 clerks, expert in the detection of fraudulent watermarks, engraving, and typography!

Sir William Congreve (1772–1828), best known for inventing the rocket as a military weapons system, tried to rescue the Bank of England. Through dint of effort, genius, knowledge, and considerable skill, he attempted to devise a method of papermaking that would outwit even the cleverest and most talented of forgers: he helped bring about the invention of colored watermarks. (Another connection between papermaking and watermarking.)

Congreve invented "triple paper" about 1818 and patented it on December 4, 1819. He made it by couching a very thin sheet of white paper on top of which he stenciled and couched one or more colored pulp designs and, while all of these were still wet, couched a third thin white sheet over the previous two. He then pressed and dried the triple sheet, which looked and felt like a single unified piece of banknote. When held to the light, though, the colored design or designs showed through.

Cost factors, among other reasons, prompted the Bank of England, on the advice of John Portals (a well-known papermaker during this period) to retain the same papermaking establishment it had long since been using for its banknotes: Portals! From 1725 to this very day, the firm of Portals has held the privilege of manufacturing paper for the Bank of England.

Curiously, Sir William Congreve requested John Portal's assistance in duplicating his triple paper

invention.

More than one claim has been put forward with regard to the invention of the chiaroscuro (light-and-dark) watermark. Despite the excellent early wire effects obtained by superb craftsmen in the Johannot paper mill (Annonay, Ardèche, France) as early as 1812, it is William Henry Smith of England who, about 1845, is said to have invented the highly complex chiaroscuro watermark.

The French claim the same invention two years earlier, but William Brewer and John Smith patented the chiaroscuro watermark in 1849 that, strangely, was acquired by the Bank of England. The reader may recall that Portals made banknote paper for the Bank of England. Isn't it interesting that Brewer, who made moulds for the Bank, worked with Congreve and Portals during the "triple-paper" fiasco? John Smith also worked at the Bank making moulds. (It would appear that we have the beginnings of a "whodunit" at hand.)

Photographic images of people and places as well as of long-revered works of art come breathtakingly to life when a chiaroscuro watermarked sheet is held to the light. All of the detail in the image is worked three-dimensionally on a wax tablet by a sculptor using tiny modeling and carving tools. On completion, the design is electroplated, and male and female dies are manufactured to impress the image upon a fine woven wire screen (48 to 60 wires to the inch). The screen is fastened to a mould, and a sheet of paper is formed in the normal manner from short-fibered pulp (for best results—at the sacrifice of sheet strength). When held to the light, the delicate modeling, the nuances of light and shade, are quite clear.

There are papermakers today (as there always will be) who are at work improving on even these inventions, and I regret not being allowed the privilege by some of these individuals to describe their contributions to the field in detail.

MAKING WATERMARKS

Some years ago, I had the privilege of visiting Henry Morris at his Bird and Bull Press. I discovered a man who gave freely of his time and procedures to a total stranger. What follows on making watermarks was shown me by Mr. Morris at that time—an example of the oldest and, perhaps, the simplest method of making a watermark to personalize your handmade paper. Because of space considerations, I will avoid a dissertation on esthetic considerations and address myself to the problem at hand: how to make our very own mark or "chop" using soft, pliable wire.

1. Make a number of thumbnail sketches of your proposed watermark. Now, discard them.

2. By this time, you have rid your mind of the first, obvious solutions; make a dozen or so more proposed designs.

3. Select a suitably sized piece of end-grain wood, and, after mulling the problem over for a while, draw one of your designs (the one that best seems to "be" you) with pencil or ballpoint pen on the surface of the block.

4. Tack headless brads along the body of your design, wherever a bend in the wire (.030 inch soft brass wire) will be required. Also affix a starting and a finishing pin, as shown, outside the main body of your design.

5. Attach the wire to your starting pin and follow along your design, bending the wire as required. Do not allow the wire to become slack. Keep going, and keep the wire flat. Now, tie the end to the finishing pin, and lift the design, gently, from the block. Solder wires that cross and any joints formed. Cut the lead and tail wires, remove excess solder from the design, and spot-solder your design to the mould face on which you will form your sheets.

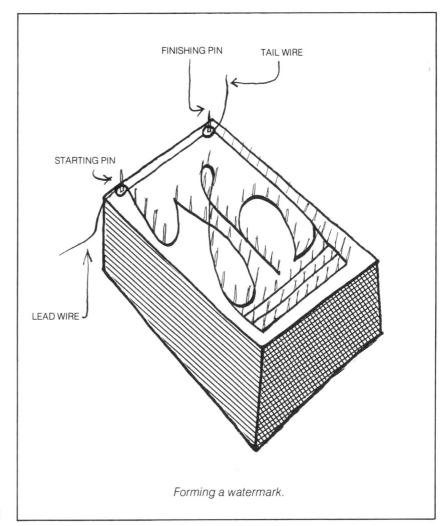

Forming a watermark.

A raised wire watermark soldered on a mould-face.

A pseudo-chiaroscuro watermark.

Above is an alternate solution to a watermark that will appear to be a light and shade watermark—yet is accomplished quite simply and without benefit of wax, electrotypes, and dies.

You will need access to printer's type, type decorations, scrap wove wire of extremely fine mesh, and a dowel rod tapered sufficiently to allow you to burnish the wove wire deeply without creating holes in the mesh.

1. Organize the type, type decorations, and the spacing between the individual units of the proposed watermark so it pleases you. Tie them together.

2. Lock the elements in a vise.

3. Place the wove wire cloth over the type, and burnish with your tapered dowel rod (or an equivalent tool). Burnish deeply. Get in *all* the crevices. Do not move the wire or create holes by stretching or piercing it.

4. Trim and shape the wire as desired: an oval, rectangle, free form.

5. Spot-solder the watermark to the mould using as little solder as is necessary.

It would not be at all remiss were you to think about *where* to place your mark or marks on any and all moulds you use. Examine the placement of watermarks in sheets you presently use and particularly admire.

Dard Hunter's chiaroscuro watermark.

This portrait chiaroscuro watermark of the late Dard Hunter was created in the traditional fashion:

1. The portrait is cut into a sheet of wax (about ¼ inch thick) by the artist, who uses a variety of small gouges and other cutting tools.

2. Working the wax on a well-illuminated surface, the sculptor removes and/or replaces wax to achieve his desired result.

3. The wax is well-coated with powdered graphite.

4. An electrotype about $1/32$ inch thick is then made, usually backed with lead.

5. A male and female electrotype may be made, and the fine screen wire (about 48 to 60 mesh) is compressed when placed between them under pressure—or a burnisher may be used with only a negative electrotype.

6. The screen is affixed to the mould and chiaroscuro watermarked sheets may now be formed.

ON DYES AND DYEING

*The hands of the dyer reek
like rotting fish and his eyes are
overcome by weariness.*
—Papyrus Anastasi

Reaching back into prehistory for some of the materials used to dye cotton or flax fibers, we find both the exotic and the unusual. Along with berries, bark, and bugs that may flourish in someone's backyard somewhere, there are boxwood, brazil wood, cochineal, Frankfort black (burnt winelees), fustic, indigo, kermes, kutch, lac, lapis lazuli, lee ashes, litmus, logwood, madder, orpiment, orseille, peat coal, red chalk, safflor, saffron, turmeric, varientia, verdigris, vermiculo, waisdo, woad, and woad ashes, to name a handfull.

It appears that few of the dyes were colorfast or bleedproof; it was necessary that the dyer grind, dissolve, and set these several colors with various solvents or mordants ("any substance which, by combining with a dyestuff to form an insoluble compound or lake, serves to produce a fixed color in a fiber. . ."). Early on, among hosts of other ingredients or materials, mordants included salt, vinegar, soda, cream of tartar, brass, iron, tin, copper, rusty nails, alum, tannic acid, and what in polite circles was termed "chamber dye" (urine).

Although papermaking, historically, has been identified as the white art, it appears as though color—the entire spectrum of color—has long since become a force to reckon with from the creamy or cold white, smooth-surfaced wove papers through textured, laid, pepper-and-salt dispersed flecks of many colors on a colored ground, to the use of ebon, blacker-than-black, matte hand or mould-made papers.

There are those among us who want no part of coloring paper, who wish to use no chemicals of any kind for any purpose, who believe in the natural uses of natural materials—and there is abundant room for such persons in the world of handmade papermaking. Problems of lightfastness, bleeding of colors, fugitive pigments, difficulties in finding the right natural dye to fit the fibers in your beater, the possible need for tannic acid or a rusty nail as a mordant to obtain the particular hue, value, or intensity you desire only serve to whet one's appetite for problem-solving.

At the other end of the spectrum, there are numbers of papermakers who seek out the most recent innovations in dye technology for use in coloring their pulp to accomplish certain ends not otherwise obtainable.

To attempt to satisfy this vast, differentiated group, the following is a brief overview of the subject. If you are interested further, consult the bibliography for expert guidance.

COLORED PIGMENTS
Finely ground, these water-insoluble colored materials act not unlike fillers and other loading materials which certain papermakers add to their beaters (such as white clay) during the beating process. To enable the fibers to retain the pigment, alum must be employed. The pigments may be divided into two separate classes:

Inorganic colored pigments. Relatively weak in hue, inorganic colored pigments possess excellent lightfast properties. Among a long list of suggested hues, I would include the cadmium yellows and reds, iron oxides, ultramarine, molybdate orange, chrome yellow, and burnt umber—to name just a few colors.

Synthetic organic pigments. The reader should exercise great caution in using these coloring materials, since they cover a great range of lightfastness and bleeding properties. Certain papermakers use high values of particular blues to tint special white papers to make them appear whiter. Color choices in this class include: Hansa yellow, Prussian blue, the phthalocyanine blues and greens, Hansa orange, and carbon black—among many others.

SOLUBLE DYES
The results of a poll of the papermakers at work today as to which kinds of dyes they use, if any, would reveal that the majority use water-soluble dyes to color the fibers in their beaters. Of the many classes of such dyes, three are the most important for papermaking purposes:

Acid dyes. Acid dyes are composed of sodium salts of comparatively low molecular weight color acids. Curiously enough, acid dyes are always alkaline because they are either sodium or related salts of color acids. Their lightfastness ranges from poor to excellent, so you will have to select colors carefully. They are, as a rule, generally duller in hue than basic dyes, but of greater intensity than direct dyes. They may bleed if wetted or moistened. These dyes are usually added *after* the sizing has been introduced and *before* the alum is worked in to the beater. (This is not a law carved on a tablet, so if you wish to reverse this procedure, feel free to do so.)

Basic dyes. It appears that basic dyes were the first synthetic dyes invented. As a class, they were brilliantly intense, but not lightfast. For reasons I do not pretend to understand, basic dyes (for those of you who will be experimenting with papermaking from wood chips and such) appear to attach themselves instantly to all the noncellulosic impurities present in your beater. It may result in paper that is more "interesting looking" than you had planned, but it also may have an acid pH that will not please you. If you want lower values of the color used, just add tannic acid to the beater.

Direct dyes. Of all of the dyes thus far mentioned, this class of dyes appears to be most relevant to the needs and desires of papermakers. As was true with acid dyes, direct dyes are alkaline salts of dye sulfonic acids. They require no alum or size to allow them to saturate the cellulose fibers with color. Yet, if you want a deeper or lower value of a given hue, it would be useful to add *both* size and alum. Similar depth of hue can be obtained, if you are unconcerned about corrosive effects, by introducing about 5 to 10% of a simple electrolyte—common salt—to the dry weight of the fiber and heating (if that is possible) the batch to 130° plus or minus 10°F.

Direct dyes should be added to the beater in a cold, dilute solution. On some occasions, you may *think* you have a dye or true solution when, in fact, you have a *colloidal* solution. Problems accompany this state of affairs, even in a 1% solution.

For optimum results, the pulp in the beater should be close to or right on a neutral pH and the dye should be added before alum (if you wish to take advantage of the alum and its effect upon the color). White clay (as was mentioned earlier) and titanium dioxide are not uncommon ingredients in the beater.

If, on the other hand, you add direct dyes to the beater in a *hot* solution, you could expect the following to happen: the first fibers to come into contact with the dye will be more deeply dyed than those in any other part of the beater, which will or could result in a mottled appearance that may or may not hold esthetic value for you.

ON SIZING

Handmade paper, after forming and drying, is called waterleaf (unsized paper). This thin sheet of interlocked, matted fibers, if it is to receive a fluid ink, paint, or other liquid upon its surface without spreading out, or feathering, requires additional treatment. If it is intended to be used in a printing press (because the inks used there are stiffer and/or tackier), waterleaf or unsized paper is preferred, as it was by the Chinese for eight centu-

ries. Waterleaf is used today by certain purists. These are personal choices.

Why does waterleaf absorb liquid like a sponge? To respond to this question, we must return to the cellulose fiber, which is an extremely hydrophilic (water loving) substance, and the fact that pulp fiber surfaces happen to contain a high specific energy.

It is believed (and again, there is no unanimity among the experts) that any liquid applied to a waterleaf surface initially wets the fibers. Then, it is absorbed into the anatomy of the sheet by the capillary action of the passages between adjacent fibers. In addition to the feathering process mentioned above, liquids also pervade the paper along its surface through cellulose-fibered paths.

Derived from the Latin term *assidere,* altered by the early Italians to *assisa,* abbreviated in later years to *sisa,* and now called *size,* the word describes both the materials and the process for making paper hydrophobic (water repellent).

One dictionary defines size as "[a]ny of various glutinous materials, as preparations of glue or flour, or varnish, shellac, or the like, used for filling the pores in the surface of paper. . . ." Looking backwards momentarily, the term sizing once encompassed so broad a definitiion as to include the introduction or addition of *any* chemical agent into the beater (or later in the tub) that was designed to modify the properties of the sheet. Today, we are interested in selected materials that will bring about water repellency without harmful effects to the paper or the use to which it will be put in future.

Some sizes are added during beating (stock sizing) and others are applied to the sheet of handmade *after* it is formed (surface sizing). Among a host of materials used as sizes are rice starch, juices of various plants, gelatine, casein, animal glue, synthetic resins, and polyvinyl alcohol. Until

Moritz Friedrich Illig (1777–1845) of Germany discovered rosin sizing (which was added to the stock in the beater) in 1807, waterleaf was made hydrophobic by dipping finished sheets into thin solutions of gelatine or glue.

Alum, added with sizing, is a shorthand term used by papermakers for aluminum sulphate and is not a true alum. It is either a coagulant or a precipitating agent used to set the size on the fibers or have it precipitated in the stock. Curiously, alum, an alkaline substance, becomes *acid* when dissolved in water, through hydrolysis. A very rough, so-called rule-of-thumb employed by papermakers in the *distant* past for setting the size was to add alum in a dry weight ratio of 1.5 (alum) to 1 (size). I am not certain it is to be trusted.

To add to the printing quality of a sheet of paper; to increase the weight of the paper; to provide opacity; to "correct" the color of the paper, especially if it is white; to heighten the brightness of a sheet; and to achieve a host of other properties, certain papermakers add numbers of materials to their beaters, including ammonium sulfamate, barium sulphate, blanc fixe, borax, calcium carbonate, calcium sulphate, casein, clay, diatomaceous earth, lithopone, silicate of soda, talc, titanium dioxide, and zinc sulphide. These materials are usually added with the sizing.

The simplest way to size paper, using the finest of materials, is as follows:

1. Soak 1½ leaves of imported leaf gelatin in 1 pint of water until the gelatine triples its original thickness.

2. Warm the mixture in a double boiler to dissolve the gelatine.

3. Pour this solution in to 3½ quarts of water. Stir well.

4. Pour into a large tray.

5. Immerse sheets thoroughly about five at a time, into the size bath. Remove.

6. Press lightly.

7. Separate sheets and dry on glass, in spurs of four or five, or by any of the other methods mentioned in this book.

It appears that Japanese papers were sized with *tororo,* also known by many other names. It is the root of a plant especially cultivated for this purpose, though it also grows wild in certain regions. In the winter, the Japanese peel the roots, beat them with water, squeeze the result through a sieve, and then add the mucilaginous extract to the vat. It is believed that *tororo* makes the paper insect-proof and retards the freeness of the sheet in forming on the mould and deckle. I wonder if it is not also responsible for allowing the Japanese papermaker to couch and press his sheets one upon the other *without* the use of felts, as is practiced in the Western world.

From most accounts, a consensus holds that the Moslem world employed, with local variations as is true about all of papermaking, sizes of starch derived from various plants, including rice, wheat, and the roots of native flora.

The use of starch as a sizing for paper in Europe is only known in the earliest phases of papermaking (up to about 1290); animal sizing was in practice as a tub size just prior to and after the fourteenth century. Derived from the ears, feet, hoofs, and assorted remains of animals sent to the slaughterhouse, the animal glue thus obtained for tub-sizing sheets was carried on until the nineteenth century. The sizing room with the stench of the materials being boiled and prepared for the sizing procedure, was also known as the "slaughterhouse," but not for the expected reasons—it was so designated because so many sheets of paper were damaged during the process. A decade or so passed before Illig's rosin sizing invention was used by European papermakers who sought a machine method of sizing.

At this juncture, we are beginning to enter the realm of proprietary interests, supertechnology, chemistry, patent rights and, for obvious reasons, let us stay with the simple and complex phenomena of the natural world around us.

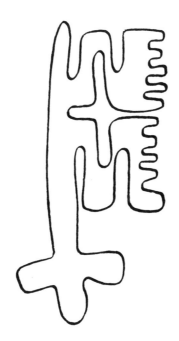

5. PAPERMAKERS AND MILLS

Here dwelt a printer and I find
That he can both print books and bind;
He wants not paper, ink, nor skill
He's owner of a paper mill.
The paper mill is here hard by
And makes good paper frequently,
But the printer, as I do here tell,
Is gone to New York to dwell.
No doubt he will lay up bags
If he can get good store of rags.
Kind friend, when thy old shift is rent
Let it to th' paper mill be sent.

—John Holme, 1696

Although it is apparent, by now, that fine-quality paper can be made from the most pedestrian of materials and tools found in the average home, it seems appropriate, somehow, to show you some of the professional equipment, tools, and materials or attitudes held by those persons who choose to make custom papers for individuals and the trade. These papers meet the most rigid standards for painters, printers, printmakers, calligraphers, designers, and others concerned with permanence of their work.

DOUGLASS HOWELL: PAPERMAKING'S OTHER LIVING NATIONAL TREASURE

I first learned of Douglass Morse Howell when I saw an exhibition of his "Papetries" in 1955 at the Betty Parsons Gallery in New York. Here were dynamic abstract collages formed on the laid wires of Howell's moulds while still in the vat! I was literally dumbfounded at the richness and variety of what I saw.

Several years later, when I was Executive Editor of *Impression,* a West coast magazine of the graphic arts that evolved into the East coast *Artist's Proof,* I received a manuscript from Doris Seidler, a printmaker, in April, 1958, titled, "Douglass Howell—Papermaker."

Key words from that article are still applicable to Douglass Howell today: "Unique, . . . A scholarly man, . . . mastery of his craft, . . . creating sculpture-like forms, . . . Book-binding and the printing of fine editions of poetry [are] a vital part of Douglass Howell's works and his papers."

Howell still uses pure linen, no chemicals, no glues, no sizing. He prefers waterleaf. He is a poet among papermakers, a philosopher among artists, an artist who seeks and makes paper that "sings," an historian, printmaker, conservator. What is he not, this father of the papermaking revolution?

His life is devoted to "seeing how beautiful a piece of paper I can make with the highest quality from

the scientific viewpoint and the highest integrity from an esthetic viewpoint." (Barbara Delatiner, "Making Paper That Sings." *The New York Times,* January 30, 1977, pp. 9 and 13.)

LAURENCE BARKER: THE RAIN IN SPAIN. . .

While Dard Hunter was, without question, the godfather of the current paper revolution and Douglass Howell the father and spiritual leader of the white art, it is to the credit of Laurence Barker, presently living in Barcelona, Spain, and making paper for artists in Europe in addition to doing his own work, that today's papermakers in the United States owe much.

Barker, when at the Cranbrook Academy of Art between 1963 and 1970, taught and influenced numbers of printmakers and papermakers who, in turn, have taught still others, who have taught others, who are teaching still another generation.

As may be verified on page 83, Barker is, himself, still an active artist who works on and with paper as his medium of expression.

His outstanding students, judging by their status in the world of papermaking today include Walter Hamady, John Koller, Aris Koutroulis, Winifred Lutz, and Roland Poska—all artists who either teach and operate their own presses or mills, or produce paper and works of art privately, or lead workshop situations, or aspects of these activities.

With regard to Barker's views on papermaking, here is his most recent statement: "For the fledgling papermaker, it often comes as a revelation that at his very first attempt to make paper he can, with a little direction, beat up a load of old cotton and form any number of sheets with no difficulty. The question arises unbidden: is that all there is to it?

"While I have now disciplined myself not to say making paper is easy, honesty compels me to observe that the learning curve is a

fast-rising one. This was brought home to me dramaticaly one day a few years ago when I was making paper for an order. My back was acting up and I felt obliged to throw my young *assistente,* totally inexperienced, into the breach. I beckoned him away from the etching press to the vat. After a few practice dips, within 15 minutes, he was forming all by himself perfectly acceptable, well-closed sheets of paper measuring 29 × 41 inches.

"Was he a papermaker? Certainly he was performing in a yeoman manner at the vat and in similarly rapid fashion he could learn the principles of couching, pressing the post of paper and, unless maximum texture was desired, the subsequent and final step of calendering or smoothing the paper.

"I intend neither to minimize nor to downgrade the skills and thoughtful attention required for every one of these functions, but the problems that arise here are nearly always mechanical in nature and consequently more easily resolved.

"To emphasize the dual nature of papermaking, it might be said with a little exaggeration that while sheets are formed at the vat, paper is made in the beater. In this sense: the only paper quality ever determined at the vat is paper thickness, adding more pulp or water for thicker or thinner sheets, whereas other paper quality results from the first and all-important steps of rag selection and beater management that might be joined together under the heading of pulp preparation. It is at this combined stage where such characteristics as color, degree of hydration, that barometer of a paper's relative hardness or softness, and fiber length including larger elements of partially beaten rag, are controlled.

"By rag selection is understood the consideration of the entire gamut of all vegetable fibers. Cotton and linen are used extensively due to their high cellulose content and exist both in textile form and in their natural state as raw cotton and flax. Other bast and leaf fibers of lower cellulose content include jute, ramie, hemp, sisal, and abaca. Synthetic fibers may also be considered, although their usefulness tends to be more circumscribed.

"Beater management is concerned with the way the rags are macerated and has an important bearing on the final paper quality. With the same ingredients quite distinct papers can be made due to different beating techniques. Not only is the progressive lowering of the beater roll and length of beating time at each setting important, but to complicate matters further, there is the notion of the sequence of feeding. By this is meant if the entire amount of raw material is not to be placed in the machine all at once, then it must be decided when each ingredient is to be added in the cycle, with the consequence that some rags are beaten longer than others, resulting in a pulp of varying fiber length and sometimes noticeably so. This can

be desirable or disastrous and never more so than when colored rags are involved. All depends upon the final effect sought.

"There is another activity that occasionally takes place in the beater which I haven't mentioned. I refer to the sizing of paper. It is true that printers generally prefer waterleaf, or unsized paper, for its maximum absorbency, but some artists who draw and paint prefer less absorbent paper. Although it is not the only way to go about it, adding size to the pulp in the beater is the simplest and most direct method.

"Returning to the matter of raw materials: so long as paper production is fairly limited, obtaining suitably soft rags is seldom a problem. When greater production is required, however, those well-worn sheets and pillow cases from friends and neighbors quickly disappear and you must turn to commercial sources.

"In my case, I found the quality of used rag available to be tougher than I was accustomed to. The rag,

in turn, produced crisper paper with some wrinkles, and this cockling was magnified and became acute on the larger sheets, so keeping paper flat, or to be more precise, making softer paper that would stay flat naturally, became a chief concern. In lithographic printing, the danger is that excessive wrinkling in the paper may cause creases during the press run, thereby ruining the proof.

"There is a way, however, to side-step neatly the vagaries of rag supply and that is to avoid textiles and raw materials altogether. The paper industry provides us with cotton linters and commercially prepared pulp that has been chemically cooked, washed and beaten. A number of very respectable papermakers use linters to the exclusion of anything else. They are, after all, 100 % cotton and given the exigencies of commercial handmade production they would appear to be the very solution, eliminating as they do the necessity of rummaging for rag with

Laurence Barker in his papermaking workshop in Barcelona, Spain. His beater is in the right foreground; he stands at his vat; his press is in the background.

the attendant chores of cleaning, sorting, and cutting. I, myself, use linters on an order of 20%, and, although I attach no particular significance to that proportion, I do not believe that linters by themselves alone can produce the finest example of the papermaker's art.

"The theme here is the paper beautiful, and what is most often prized from an esthetic standpoint in a sheet of handmade paper—texture and deckle edges aside—is that animating quality, however subtly informed, that bespeaks the paper's origin, namely the rag and fiber from which it comes. Nor do I refer to fantasy papers, those veritable extravaganzas of the papermaker's art, alive with writhing threads, but rather to a quieter, more decorous region where yet there exists an awareness of fiber presence.

"Linters are clean and useful in the sense of their immediate availability, but they are overrefined requiring but minimally additional beating for the vat. A more serious objection to them is that they represent a substitute for involvement and block all avenues to investigation and discovery. Papermaking, it seems to me, ideally begins as an adventure with rags and fibers and for the inquisitive-minded long after the mechanical skills have become honed and routinized, this transmutation of fiber in its most inclusive sense to pulp will continue to fascinate, for it is here, at this stage, where beautiful paper is conceived."

THE WILFERS: UPPER U.S. PAPER MILL

Joseph Wilfer is the friendliest friend handmade paper ever had. He founded, almost singlehandedly, the North American group called the Friends of Paper, and organized the First North American International Handmade Paper Conference in Appleton, Wisconsin, in November, 1975.

Wearing one of his numerous exotic hats, Wilfer is the proprietor, along with his brother, Michael, of

(Top) Upper U.S. Paper Mill. Courtesy of Joseph Wilfer. (Above) Portrait of the Upper U.S. Paper Mill looking west. Starting at the upper left and moving clockwise, there are two beaters, two standing presses, a mould and deckle, the edge of a vat, a bucket of pulp in the foreground, and a pile of felts.

the Upper U.S. Paper Mill, which has established an enviable reputation among professional art circles on both the East and West coasts. "In the late 1960s, while a printmaking student at he University of Wisconsin at Madison, I became interested in papermaking. By 1969, I had gathered the necessary equipment to transform rags into pulp and pulp into paper. For five years, work was carried out in a small showerroom in the basement of the Madison Art Center. In 1974, the mill took on a new name and a new home in a converted dairy barn outside of Oregon, Wisconsin.

"The Upper U.S. Paper Mill was established as a resource center and a collaborative workshop where artists could create unique pieces in and of paper. My early interest in paper was one of curiosity and experimentation, but I soon realized the inexhaustible potential of paper as a medium. Think of it. In the 500 years since paper has been known to western civilization, artists have been limited to working on what was available: rectangular sheets of paper which, over the years, became purer, whiter, and more uniform. For the first time in history, artists can work with paper as a medium, not merely as surface. They can create any size, shape, thickness, or texture they may desire. They can think of a total piece, and the paper can be more than a traditional carrier of ideas—it can be an integral part of the work. This has been the most significant motivation and major concern of the mill.

"My approach to papermaking has never been scientific. I do not want to know the reasons why. I am content to accept the magic of the medium. I have never dreamed of manufacturing pristine, quality sheets of paper. The harder I tried, the more machinelike they became. My esthetic is grounded in the imperfection, the varied, irregular beauty which makes each piece a unique entity. I visited many commercial mills in the United States,

and, in 1974, traveled to England to see large-scale hand operations at Hayle Mill and St. Cuthbert. I studied the people, the equipment, and the facilities that were required for a quality hand operation to become a profitable venture. I decided, then, that I was more interested in being a papermaker than a business man.

"Out of necessity and design, the scope of the Upper U.S. Paper Mill is limited to working in a one-on-one relationship. A handshake, rather than a contract, is the order of business and experimentation replaces tradition. It is my hope that this looseness of operation will encourage the creation of unique and significant works of art in this old/new medium.

"Regarding advice or formulas for making paper, I can only recommend experimentation, testing, and learning by doing. There are no real secrets. You can obtain the basic information from text books or can spend years in school learning the chemical and physical properties of paper science. In the meantime, you will miss all the excitement and fun of actually making paper. You might keep in mind that soft, lintlike blotter paper and hard, translucent glassines are all made from the same raw material. The basic difference is in how the pulp is prepared. Variations in the types of equipment, raw materials, water, etc., all affect procedures. Try it one way; keep a log or journal; and keep testing until it makes sense to you and you understand what works best. If I could learn to make paper, anyone can. It's common sense, interest, and stick-to-itiveness."

DONALD FARNSWORTH
Proprietor of the Farnsworth Paper Mill, Donald Farnsworth has, in a brief period, become one of the most significant, scientifically knowledgeable, best-grounded, keenly sensitive persons working today in the field of papermaking, especially on the West coast of the United States. His myriad collabo-

rative efforts with artists of many persuasions testifies to a well-deserved reputation.

"My interest in making paper by hand originally was to produce paper which would satisfy the special requirements of art conservation. As professional printmaker as well as art conservator, I experienced firsthand the range of improperly beaten commercial papers, made from the lint of cottons—too soft, too weak, too acidic and filled with wood fibers. Inevitably, the requirements for saving endangered works of art on paper implied the ideal of creating permanent, fine paper for artists and printmakers to use from the very beginning of their creative activity. Out of this need, the Farnsworth Paper Mill was started in Oakland, California principally to manufacture paper for art and conservation. In recent years, however, artists themselves have applied such vitality and invention in extending the properties of paper to the solution of esthetic problems and lending expression to their ideas that the Farnsworth Paper Mill has developed into a studio for experimental art in handmade paper.

"As a vehicle for drawings, paintings, manuscripts and books, fine paper has been prized by artists and artisans for 2000 years. The history of paper is attended by a powerful mythology such that until the present, handmade paper has seldom been used by artists in innovative ways. Computer banks information exists and are available on paper and its manufacture, and yet a negligible amount of research has concerned itself with handmade paper as art rather than a support for it.

"Because of, or perhaps as a reaction to, more than a generation of industrialized art—art concerned with the very materials of art, and especially art extending from impermanent, even self-destructive materials and processes —a considerable number of artists are rediscovering in handmade paper a plastic, sensual, expres-

sive, and permanent artistic medium. Now these artists all over the country—some at the Farnsworth Paper Mill—are applying original thought to this traditional old material and extending paper in novel directions. Whatever contributions the mill may have made jointly with these stimulating artists, conservators, and artisans have been offered with the greatest pleasure in collaboration and admiration for their achievements.

"More and more universities and individuals have requested my assistance and advice in setting up a limited facility for producing handmade paper art. To those people who have surmounted the problems of locating and using a beater, mould, and press, I suggest in general to use high quality (new) rag fibers, pure water (no iron salts), and slow beating. An-

other recommendation is to add a chemical buffering agent to the pulp at the beginning of the beating process. Magnesium carbonate and calcium carbonate are two common chemicals used for this purpose. Both, when used at about a 3% suspension, will neutralize the stock and act as a buffering agent against acid. Internal sizing, such as Aquapel and Hercon 40, are more effective in an alkaline system."

O HANDMADE PAPER: ANDREW JAMES SMITH

Andrew James Smith (b. 1945) is a unique phenomenon in the small but growing world of handmade papermaking in North America. He "feels he can boast of having 'the largest handmade paper facility in the western hemisphere.'" (Greg Gatenby, "A Creative Approach to

Don Farnsworth at the vat, removing the deckle from a newly formed sheet; David Kimball is couching a sheet in the background.

Papermaking." *Quill & Quire Update*, Vol. 42, No. 12. September, 1976, p. 3.) How many individuals own a 50-pound Hollander-beater and all the necessary other equipment housed in a huge, foam-insulated barn in Beaverton, Ontario?

Essentially self-taught—though he consulted experts including J. N. Poyser, of the Canadian Pulp and Paper Institute—Smith is an artist whose medium is paper pulp, who "has invented a full-color, half-tone watermaking technique that he claims is so efficient it would have allowed *Life* magazine to have been published without ever going to the printer."

In addition to running a commercial handmade paper mill (O Handmade Paper), which would exhaust most individuals, Smith finds time to exhibit his personal paper works and to muse on private points of view about art: "One day I realized that oil painters for 2,000 years have been all wrong, using the wrong part of the flax plant. They should use the cellulose from the secondary layer of the stem instead of the linseed oil of the flax plant. Oil is glossy, acidic, corrosive; it has no longevity, and its nature alienates the artist from his environment."

See Beverly Plummer's statement on page 163 and compare her views with those of Smith: "To me, painting is a form of papermaking because both start with a violent act—grinding, beating senseless a vegetable material, and if a painter doesn't beat his materials, then he'll sublimate this violence and it'll come out in his work subconsciously."

HMP: THE WORKSHOP OF JOHN AND KATHLEEN KOLLER

Artist and papermaker, John Koller, has been of enormous help to paper enthusiasts. The quality of papers created by HMP have made it envied by many throughout the world—including Koller's teacher, Laurence Baker.

"I find paper of continuing interest because of its simplicity. It can

(Top) Kathleen Koller at work finishing a custom work. Photo Betty Fiske.
(Above) John Koller–stripping newly made sheets from the post of felts, after the initial pressing. Photo Catherine Luyster and Brian Swift.

be made by most anyone under very crude conditions with wonderfully expressive and even sophisticated results. Made of a common plant material, cellulose, often from the wastes of other processes, paper is exceptionally tough and receptive. It has prospered in the East and the West and managed to reflect each culture. And, perhaps most noteworthy of paper's qualities is what might be called its humility—in the sense of quiet service. This quality seems in some danger of being swallowed up in the current rush. While I sympathize with the desire to better understand and use paper and the papermaking process, the urge to isolate and pedestalize this material may be a disservice to it."

TWINROCKER, INC.
I am delighted to include a portion of a statement by Kathryn Clark of *Twinrocker* on papermaking:

"Howard and I have always held similar attitudes toward papermaking as a handcraft and our particular involvement in it. First, we have a tremendous regard for craftsmanship and the stability inherent in tradition. A fine sheet of handmade paper is timeless and anonymous. It is like the bud of a flower with a life of its own. Second, we feel that handmade paper should be what machine-made paper cannot be. Making one sheet at a time allows for a greater ability to change the esthetic and technical characteristics of a paper before making a commitment to the edition. Because making paper by hand is inherently a slower but more flexible process than by machine, it permits shorter runs and innovation that would be unthinkable on a machine. Consequently, we can afford to attempt making a paper that may not have been made before, sometimes creating a paper that is quite unusual and certainly far from traditional.

"Attitudes toward paper have changed drastically in the last several years, even more so since my first involvement in 1967. Openness to new ideas in this country and the lack of tradition in hand papermaking have allowed these new attitudes to expand and develop. It is my hope that the current surge of interest in papermaking will grow through knowledge and experience and that a balance of tradition, craftsmanship, and artistic innovation can be reached."

IMAGO PAPER MILL
In addition to offering workshops to paper enthusiasts, selling paper, pulp, and other materials to those who do not have access to a beater, it appears that Robert L. Serpa (a student of Don Farnsworth) and Tim Payne are attempting to launch Imago Paper Mill as a fine art, book, and writing paper mill:

"It is our intention to offer a domestic source for these papers that will allow us to be competitive with the world market. To this end, we have obtained many new tools, the most impressive of which is our double ram 3×4 foot press. Splendid machine! No more than 25/1000 deflection at the outer extremes of the platen under full capacity. Our production moulds are 17×21 antique laid—which has a foolscap deckle, as well as a regular deckle, 18×24 antique laid and 22×30 wove. We have an assortment of shop (home) constructed wove moulds for our workshops; and, we were able to obtain enough new woolen felt material to make 150 production felts."

WOOKEY HOLE CAVES LTD.
Visitors to Wookey Hole Cave and Mill may witness demonstrations of

Howard and Kathryn Clark examining sheets of handmade paper at Twinrocker.

Tim Payne and Robert L. Serpa at their double ram press.

handmade paper being made daily by John Sweetman and his staff in the old mill building, which is presently part of an industrial museum in Wookey Hole, Wells, Somerset, England. It is believed that there was a mill on this site as early as the year 1065!

There are hundreds of old hand moulds for the tourist to see that bear a great variety of watermarks for legal and other documents.

Mr. Sweetman, in a recent letter, is equally astonished by the paper-making revolution:

"[I]t is quite amazing just how much interest there is at the moment in making paper by hand, after so many years of neglect, with only a few voices crying in the wilderness. Besides Green and us there is a Mr. Partridge who is working in an old paper mill in Lancashire."

Indeed, it is truly amazing. Only three handmade paper mills left in all of England!

THE TWO RIVERS PAPER CO.

Directed by R.S. Partridge, Managing, and C.R. and C.R.S. Partridge, Directors, the Two Rivers Paper Company at Rosebank Mill, Stubbins, Nr. Bury, in Lancashire, England produces 90, 140, and 200 pound handmade watercolor paper in white and four tints, as well as in two finishes.

"Our present largest sheet size is Demy or 18⅓ × 22½ inches. We are currently installing plant for making the metric A2, for which the untrimmed sheet will be approximately 17¾ × 25¼ inches."

BARCHAM GREEN AND CO., LTD.

Simon Barcham Green recently reorganized Hayle Mill, Maidstone, Kent, England to include Terence Franks, Graham Clarke ARCA, and himself as Directors. Their stationery reads, "handmade papermakers since 1805."

Most printmakers and painters need little or no introduction to the quality of J.B. Green papers— there appeared to be a paper for every artist's need.

With reference to my query, Mr. Simon B. Green replied:

"Thank you for your letter of 18 May regarding the book you are compiling on handmade paper. We are not really involved in the current fashion of making pulp sculpture as such, being manufacturers of handmade papers since 1805, although from time to time we do do interesting experiments of this nature for our own amusement.

"I am enclosing one photograph of our vatman at work. I would also refer you to an article I wrote for *Fine Print* just over a year ago."

Vatman at work, Barcham Green & Co., Ltd.

PART II. THEORY AND SOME PRACTICE

6.
VARIATIONS IN PAPERMAKING

Those pieces of rag be quick and bring!
The dusty old shreds are just the thing—
For pulp, for pulp, to record life's wrong,
For pulp, for pulp, for a poet's song.
It comes out smooth, and glossy, and thin,
From rollers, and wheels, and cylinder's din,
For lords and ladies their notes to indite;
For petty poets, who scrawl by night.
And newspaper scribblers who bluster and blow;
For little love-letters where compliments grow;
And stories in which the afflictions of men
Are wretchedly told by an unskillful pen.
On just such rags as once wiped away
The tears whereat thou weepest to-day.

—Carmen Sylva, 1889 (pen name of Elizabeth, Queen of Rumania)

At this juncture, it should be clear that there are many ways of making paper, that artists have widely differing views as to what constitutes the proper kind and quality of paper for whatever reason the artist wants paper, that each specific medium, such as printmaking, watecolor painting, drawing, printing (for books), and three-dimensional work, demands not one but many different kinds of paper to meet these several needs. That the same stock, using different additives to the beater in different time sequences, different beating times, different space relationships between the beater roll and the bar, and so on will produce different papers.

Unfortunately, proprietary interests, closely guarded secrets of production, patent rights, and the investment—on the part of certain interests not at all concerned with artists making handmade paper for their own specific reasons—of millions of dollars delimit what information can be shared. If I appear to generalize or waffle about specific formulas, you would be entirely correct in assuming that I have traveled but so far in my quest to expose all secrets to the reader and have, in some instances, failed miserably.

PRINTMAKING PAPERS

In printmaking, there are but four basic processes employed, each of which demands its own paper requirements: the relief process (including flexography), the planographic process—popularly known as direct or offset lithography—the intaglio or gravure process, and the stencil process, or screen printing.

In point of fact, there is no paper that is suitable for all or some printing processes all or even some of the time. There are too many variables affecting its properties as to make each paper more or less moisture-resistant in a given situation at a given time. The paper may be more or less resistant to distortion, or exhibit good, bad, or seemingly indifferent receptivity to ink; it may or may not reveal tendencies to curl, shrink, or stretch; hopefully, it will be free from dust and lint and will have good ink-drying qualities. Further, one would expect that paper for printmakers would reveal good formation and have quite high pick resistance.

Waterleaf paper (nonsized paper) is, as we know, a web of interlocked fibers that may or may not suit every printing or printmaking requirement because it does not present a fully continuous, homogeneous surface. Thus, in certain ways, we can improve smoothness and gloss, ink absorption, and the multitude of printing characteristics necessary to printing and printmaking papers by coating and/or sizing our papers.

Variable atmospheric conditions keep altering the paper. After all, it lives, it breathes and, to a man like Douglass Howell, it sings.

RELIEF PRINTING PAPERS

Relief printing may be defined as the process by which a raised, inked surface, through direct pressure, transfers the design or page of type to paper. Or, the printing from a type-high wooden surface of that which remains after the printmaker cuts away the lines and areas he does not wish to see printed.

Whether one uses an old Washington hand press, a platen or flatbed-cylinder press, a rotary letter press with its wraparound, shallow relief plates, or a tablespoon or baren (a Japanese burnisher) with one-person hand pressure, the problem, more or less, is the same as stated at the outset.

The paper must be smooth, even, or level for best results. It may be gossamer-light, airy, and strong as are Eishirō Abe's *gampi* papers, or it may range the gamut of very soft to very tough Japanese and other papers of all thicknesses, surfaces, textures, and colors—if one is using the traditional baren or tablespoon. It may be one-sided or coated, which suggests it is two-ply: the original waterleaf stock and a layer of mineral pigment—such as blanc fixe, calcium carbonate, china clay, stain white, and titanium dioxide—bonded with an adhesive (casein, a vegetable protein, or a proprietary mixture) to the stock. It may be two-sided (which suggests that two coated sides sandwich the original stock. Or it may be supercalendered or enamel stock, depending upon the press employed and whether finescreen halftones were used, as well as other considerations.

Paper for relief printing should be receptive to ink and sufficiently resilient so it assumes its original form for successive color printings from blocks and/or type or halftones.

INTAGLIO PRINTING PAPERS

In this instance, the image to be printed exists below the surface of the plate or cylinder. Whether an etching press, photogravure (single-sheet-fed), or rotogravure (roll-fed) press is used, the same problem exists of hand-wiping the surface of the plate, leaving the ink down in the lines or pits or, in industry, having a doctor blade perform the same funcion on the rotating, pitted surface.

Paper—for intaglio or gravure printing—should be on the soft side, relatively smooth, and absolutely level without any surface irregularities. It, too, should be resilient and of uniform quality throughout; it should be absorbent and strong to the degree that you can soak each sheet in water thoroughly, blot dry the surface of the limp sheet, and print the intaglio plate under powerful pressure without bursting the paper or having the sheet cut by the plate.

In industry, gravure inks are more liquid (they bear little resemblance to the thick, viscous inks employed by artist-etchers) than inks used with "old-fashioned" hand methods and materials. Thus, they are more quickly absorbed by the paper. The faster these indus-

(Right) Close-up (50x) of a waterleaf sheet. (Below) Close-up (50x) of a handmade drawing paper. (Opposite Page Top) Close-up (50x) of a handmade printmaking paper. (Opposite Page Bottom) Close-up of a handmade watercolor paper. All photos C. Antonie.

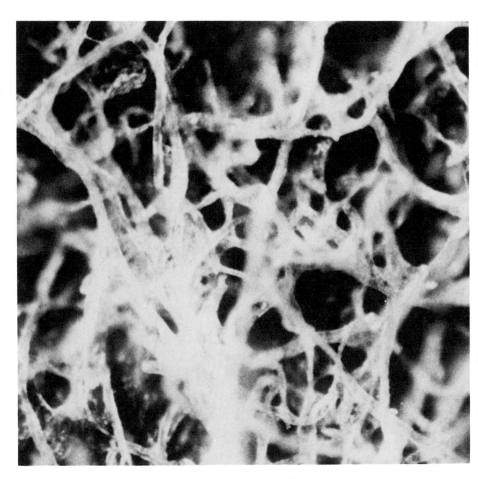

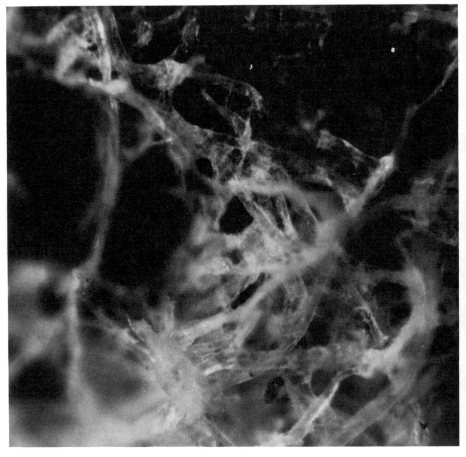

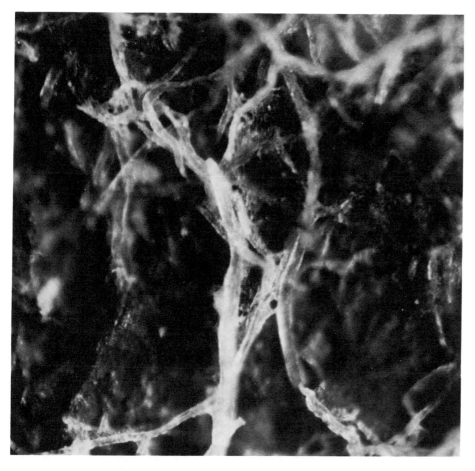

trial presses run, especially roll-fed, super high-speed equipment, the more liquid the ink employed. These machines have various methods, such as heat, for quick-drying inks used in gravure printing.

PLANOGRAPHIC PRINTING PAPERS

The image in direct or offset lithography (a planographic process), by definition, exists on the surface of the stone or plate. Relying on the principle that grease and water do not mix, artists draw with a greasy substance on a limestone or zinc or aluminum plate (or photographers shoot the image on a photosensitized plate or stone), which is then chemically treated, kept wet with water throughout the printing process, and rolled up with a greasy ink (which adheres to the undrawn areas). All of this places certain demands upon the paper required for best results.

The only major difference between direct and offset lithography lies in the fact that the sheet of paper in the former approach, is placed *directly* upon the inked stone and then printed, while the inked image in the latter approach is printed first on a rubber blanket and then transferred to paper.

Since much water is used in both methods, lithographic paper demands great internal and surface strength (probably obtained through additives and sizing—perhaps in the beater *and* after the sheet is formed). It also should be flat, even, and fairly hard. In offset printing, however, papers of various rough-textured surfaces may be printed upon.

There is little point in attempting to describe the high-speed, web, offset-lithographic machines that print layer upon layer of colored inks simultaneously (or within seconds of each other, because of superspeed, ink-drying capacities) on both sides of the roll of paper as it flies through these huge machines. You will probably not be making paper for such machines.

SCREEN PRINTING PAPERS

There are few, if any, special requirements for screen printing, since the process is equally adaptable for use on plastic, glass, fabrics, metal, and so on. It can even be adapted to print on complex three-dimensional forms through the use of specially prepared jigs (wooden devices for special three-dimensional printing techniques).

Screen printing is a stencil process. A metal wove screen or a synthetic fiber of very fine wove is stretched tautly across a frame which, in hand methods, is hinged to a baseboard. The image is created, one stencil at a time, by blocking out areas not to be printed, placing a sheet of paper under the screen guided by predetermined registry marks, and squeegeeing the viscous ink across the screen down on to the paper. When the first run or color is dry, the process is repeated with various colors and new stencils. The sheet of paper is always placed in the same registry marks.

In commercial work, as in plastic and fabric printing, web screening may be employed, as well as sheet-fed screen printing, on various kinds of presses adapted for this venture.

For best results, paper for screen printing should be stiff, bulky, of reasonable dimensional stability, and heavy in weight.

ALL-PURPOSE PAPERS

If you are a printmaker, I would strongly suggest that you save the scraps of the several papers you use for the particular printmaking medium in which you specialize. No doubt, through the years, you have found through trial and error that certain papers appear to work much better than others. You have probably also found that you must, from time to time, trim various of these hard-won papers for myriad reasons.

So. Tear these redundant scraps into 1 to 2 inch squares, and soak them in a plastic garbage pail. Keep papers for each printmaking process separate, at first. Later on, you may wish to combine the scraps together to discover what happens . . . as did Dr. Schaffer in the eighteenth century.

Allow the paper scraps to soak for a day or so while you refresh your memory by rereading Chapter 2. Then, follow the directions below.

HOVER'S SIZES

After numbers of experiments, this formula was developed by a Mr. Hover in 1867 (see Davis, pp. 434–435, in the Bibliography):

1. Dissolve 4 ounces starch in 240 ounces of water.

2. Add 12 ounces of commercial carbonate of lime, magnesia, or its equivalent.

3. If desired, add a small quantity of glue (for sizing).

4. Size, dry, and calender the paper.

The ingredients listed above may also be added during the beating process if you prefer; the desired result is to have the paper coated with carbonate of lime, magnesia, or their equivalents.

Two years later, Mr. Hover patented the following recipe for similar purposes, claiming it both made

---

A Note from Imago

I mago Paper Mill, through proprietor Robert Serpa, recently wrote: "I would be happy to pass on 'recipes' for various papers, but I am sure that my formulas will be beneficial only to my process. That is, the paper you produce is dependent on machinery you are using. I could take my proportions and beating times and, more than likely, I would not be able to produce the same paper with your Noble & Wood. Each tool has to be approached from its unique style of beating. For example, when we used a Noble & Wood in this shop, we were not able to beat with full pressure for very long periods because the blades would cut the fibers very quickly. Therefore, we would brush the fibers lightly to get a nicely hydrated pulp. On the other hand, the blades in the Valley beater are ground in such a manner as to allow full beat hydration without excessive cutting action.

"In general, for litho and etching papers, one would use a linter fiber base that is beaten so as to allow fairly rapid draining potential when formed on the mold. Staple fibers are added to improve the strength of the paper to overcome plate cracking problems on the printer's end. The exact ratio of linter to staple fibers would be determined by the individual beater's action. If your beater cut the staple fibers too easily, there would be little advantage in using them. Trial and error becomes the order of the day. For the sake of argument, I would recommend a 75 to 25% (linter/staple) fiber content to start your experiments.

"On the use of additives for more art-oriented purposes, you can use methyl cellulose to stiffen the paper. Additionally, the use of CMC (carboxymethyl cellulose) allows the artist to build large areas of relief on flatter surfaces. The CMC changes the viscosity of the water and suspends the fibers in such a manner as to allow you to fill a turkey baster and apply the pulp in layers. As the pulp dries, it will shrink, but the amount of pulp needed to produce large areas of relief is markedly reduced. The problem with CMC is that it takes a lot longer for the material to drain when used on a mold, but it does give the paper a nice snap when it is cured."

the paper whiter and improved its surface for printing:

1. Mix together 7 gallons of "ordinary glue sizing" and 1 gallon of acetate of lime.

2. Dry and calender the paper.

WATERCOLOR PAPERS

The watercolor painter interested in making handmade paper has at least two options: either refer to Chapter 7 to learn how simple it is to recycle trimmings and scraps of good quality, pH neutral papers or reread Chapter 3 on the "simplest method of making paper" and add the following steps *after* you have removed the cheesecloth from the moist sheet:

1. Place the moist, felted sheet that still bears the imprint of the cheesecloth (which certain watercolor painters may find objectionable) on a larger piece of rough burlap or other rough-textured fabric, such as gunny sacking.

2. Place another piece (same size) of the same rough-textured fabric over the moist sheet.

3. Handling the "sandwich" of gunny sacking (or its equivalent) and the moist sheet of paper carefully, run it through a clothes wringer or an etching press with sufficient pressure to imprint the unique texture sought by many watercolorists on the fabric.

4. Size the sheet with a large clean, housepainter's brush (the largest you own or can buy) with a starch or gelatine sizing—or any other sizing that suits your fancy (see pages 80–81).

DRAWING AND CALLIGRAPHY PAPERS

There is no particular sheet of paper that will satisfy the definition of "drawing paper" or paper suitable for the art of calligraphy by all artists.

There are many artists who, when thinking of drawing paper, instantly imagine the thin, strong, laid paper of French origin that was tra-

---

Don Farnsworth's Sculpture Pulp

*U*sing his five-pound Noble & Wood Hollander Beater, Don Farnsworth (b. 1951) was kind enough to share this formula for sculpture pulp:

"Through my paper conservation background, I learned about methyl cellulose, and it has been very valuable to me in many ways. It is up to archival standards—neutral pH, indefinite shelf life, impervious to fungal and insect attack, etc. . . . (See Suppliers List.) I mix ½ pound of this cellulose to 5 gallons of water, and keep this paste around for various uses, such as sculpture pulp. To make the methyl cellulose pulp:

1. Tear 3½ sheets cotton linter PS21; add to 15 gallons of water in the beater.

2. Put blades in up position (clearing position); whip for 15 minutes.

3. Add 450 milliliters of methyl cellulose during the first 5 minutes.

4. Add between 35 and 40 milliliters Aquapel during the last 2 minutes. Add it slowly, diluted with water. (Since Aquapel must be purchased in 55 gallon minimum orders, an excellent substitute—recommended by Mr. Farnsworth—is Hercon 40, which is sold in one gallon containers.)

---

ditionally used as the ground for a charcoal drawing; others conjure up a coated board or paper such as Bristol or illustration board; still others think in terms of the vast choices of Japanese papers available on the market; and there is a considerable number who disagree with all of the above.

HMP PAPER

I am indebted to John Koller for sharing this professional, proprietary formula with my readers. At the same time, it should be stated that without his custom-made equipment, materials, tools, beater, water supply, etc., etc., you will probably not obtain the same results as this master artist-papermaker. Custom sheets of handmade paper are not arrived at without knowledge, many years of experience, skill, labor, love, equipment, pure water, and a host of other requisites.

Here is Mr. Koller's shorthand description of his HMP, a warm white, cold-press, 90 pound, 17½ × 23 inch lightly sized multipurpose drawing and/or printing paper:

1. Pulp: 25% well-beaten (the consistency of cream) old linen rag. 75% lightly beaten linter.

2. Formation: add 3 quarts of pulp per sheet.

3. Pressing: to 150 pounds per square inch (between felts). Second pressing (while still wet but without felts) to 75 pounds per square inch.

4. Drying: allow to harden for 2 to 3 weeks.

5. Sizing: use rabbit skin glue mixed in a ratio of 1 pound per 6 gallons of water. Dip the dry sheets into the warm glue bath and keep them submerged until they are thoroughly wetted. (Dip in groups of 5 to 10 sheets).

6. Stacking: stack the sized sheets without felts, and press to 75 pounds per square inch.

7. Drying: heat should not be greater than 90°F.

8. Finishing: press the thoroughly dry sheets to 150 pounds per square inch for 8 hours.

John Babcock's Papermaking Methods

*U*sing his recycled washing maching/hydropulper, John Babcock demonstrates his process at left. Note the two baffles glued and screwed to the inside of the ex-washing machine. This is to prevent the pulp from embedding the operator from head to foot in linters within seconds of turning on the switch controlling a high-speed impeller in a circular vat. Babcock is using Hercules linters PS21—he uses 2 full linters for sculpture pulp, and 1 or 1½ linters to approximately 10 gallons of water for forming sheets of paper.

The photograph provides some sense of the speed with which this recycled machine works. His powdered pigments (earth colors, primarily) come from Fezandie and Sperrle (see Suppliers List). Babcock adds tiny amounts of Reten 210 and Aquapel from Hercules, Inc. In some works, he adds only Hercon 40.

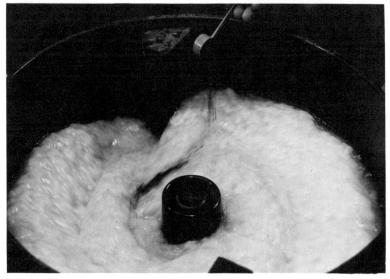

(Top) Adding a piece of linter to the whirling pulp. (Above) Adding powdered pigment to the swirling pulp. Photographs Richard Kluge.

PULP FOR THREE-DIMENSIONAL WORK

If paper relief casting for printmaking or sculpture is your particular bent, the problem of preparing or purchasing pulp is the simplest problem we have, thus far, encountered.

If you have no facilities at all and little desire to indulge in the process of making pulp for relief work, please consult with the Suppliers List and select a manufacturer who can supply your needs in quantities and prices that can accommodate you. In certain instances, it may be necessary to find a number of colleagues equally interested in purchasing pulp to meet a minimum order requirement. Whatever, all of the sources listed will provide pH neutral pulp suitable for your work.

For those who wish to start from scratch, it is suggested that you make as much pulp as you need in one session from cotton linters, using one of the numbers of methods previously described. Here is an instance when a hydropulper of sufficient capacity to satisfy your requirements would be a most useful tool, and one that would perform the task in short order.

Pulp for three-dimensional work need not be beaten as long as paper for sheets; need not require the care and concern at every step along the way; and need not demand the variety of tools and equipment. This is not to suggest that I am deprecating, in any way, the users of three-dimensional approaches or the material used. As I have repeatedly said, "There is, to my way of thinking, no one way to get to heaven in the arts."

MISCELLANEOUS SUGGESTIONS

In no special order of importance, I have listed a number of vague—because they were or are proprietary—commercial processes from various sources which may be of some use to those who use paper, make paper sheets, or view paper as a medium of its own and wish to experiment and/or play with these "paper bullets of the brain." See the Bibliography for greater detail.

Abrasive papers or textured areas. Screen glue on to a strong backing paper, such as kraft or hemp. Powder fine sand, carborundum, emery dust, garnet, mica specks, flocking (to name but a few things that first come to mind) on to the gluey surface. Shake off the excess. Allow to dry.

Deckles. Why not circles, ovals, the model for an envelope, a heart, any irregularly shaped deckle? Substitute your newly shaped deckle for the normally used rectangular one.

Thick mats of paper. Through multiple couching, one sheet upon the other, you can build up layers of paper, gauze, asphalt (maybe), more gauze, more paper, and so on to a thickness of inches, if desired. Useful for embossed prints (see Michel Ponce de Leon, in the color gallery) or even a doormat, if you varnish the final layer.

Pulp drawing. Draw directly on gauze with a turkey baster filled with pulp, and felt the drawing.

Colored pulp drawing. Use various colored pulps (one at a time) in a turkey baster. You may have to cut a piece off the tip of the baster, if your pulp is too thick. Draw your image directly on to the screen surface of the mould. Allow the drawing to drain thoroughly. Now, dip the mould and deckle into a vat of white or other pulp and follow the usual procedures in making a sheet of paper. In this case, a unique sheet of paper with a multicolored pulp drawing as part of the sheet results.

Translucent glassine. For those of you interested in making translucent glassine papers, reread the description by Winifred Lutz on page 101.

Impregnated papers. These are papers that have been painted with, dipped, or soaked in sundry solutions to make them transparent —brushed with a thin coat of a so-lution of Canada balsam in turpentine; waterproof—either coated with liquid paraffin or soaked in shellac; greaseproof—a translucent glassine sheet immersed in a solution of salt and albumen; fireproof—sprayed with a fireproofing chemical or painted with acetyl cellulose, which is soluble in acetone.

Metallic point drawing paper. Want to draw with a pewter point? Size the finished sheet with one or more thin coats of lime, whiting, and zinc white. For aluminum, brass, copper, lead, silver, etc., the coatings may be a combination of barytes, clay, blanc fixe, lime, and whiting mixed with gelatine. Then draw.

Fake mother-of-pearl paper. Using a coated stock, float the sheet on a solution of salts of lead, silver, or bismuth. Lift slowly, and allow to dry. Expose the sheet to vapors of sulphide of hydrogen. Then, pour

---

Harrison Elliott's Traditional Approach

*H*arrison Elliott once had a one-man paper mill on the seventh floor of what was then known as the Stevens-Nelson Paper Co. and another at the Merchant Marine Rest Center at Gladstone, New Jersey, which accommodated the entire process of handmade papermaking. Here is how he made paper.

1. Cut cotton or linen rags into tiny pieces to prepare them for the beater. If you are going to use a blender, the rags must be small— about ½ inch squares—and soaked for a day or two.

2. Beat, blend, or mash the rags with a mortar and pestle (take your pick) until, when a sample is tested in a small glass, there is no further evidence of what once was cloth. Just a fibrous, liquid mass. If you use a blender, place only a tiny amount in the container with a great deal of water. Watch for telltale signs of blender motor distress, and turn the machine off instantly. Use less rags, more water, and turn the blender on again—in short bursts.

3. Make enough pulp batches for the paper you want. In a vat filled with water, add pulp until you have the proper consistency.

4. Dip your mould and deckle into the vat, and give it the vatman's stroke as the water drains from the mould.

5. Remove the deckle, place the mould pulp-side down on a dampened felt, and couch the sheet on to the felt. Lift the mould carefully.

6. Repeat Steps 4 and 5 until you have formed a post, or pile, of wet sheet, felt, wet sheet, felt, and so on. Squeeze the post in a bookbinder's press. Then, separate the sheets from the felts. Repress the semi-dry sheets to remove the felt marks, and spread the sheets to air-dry them in spurs of four or five.

7. When the sheets feel dry, they are ready to be sized with household gelatin size. Heat a mild mixture of gelatin size in a pan larger than your sheets of paper. Dip each sheet in the hot bath, and put all the sheets back into the press for a further squeeze.

8. Remove the sheets, carefully. Air-dry. Place them between sheets of zinc, two at a time, and once again, submit the sheets to pressure. Thus your sheets will obtain that certain gloss prized by certain users of handmade.

collodion over the surface of the sheet and witness unique color phenomena.

Parchmentlike paper. Take a sheet of waterleaf made of rag. Place it in a diluted bath of sulphuric acid. Remove with wooden tongs. Wash thoroughly with water. Dry.

Iridescent paper. "Boil 4½ ounces of coarsely powdered gallnuts, 2¾ ounces of sulphate of iron, ½ ounce of sulphate of indigo, and 12 grains of gum Arabic; strain through a cloth, brush the paper with the liquor, and expose it

quickly to ammoniacal vapors." Davis, p. 584.

Woven paper. Twist paper (which has been treated with neoprene) and weave desired forms. One can even make washable (with soap), wear-resistant rugs or floor coverings in this manner.

Enameled writing surfaces. The first coat is comprised of a 10 % solution of Kremnitz white, glue, and water in which is dissolved white shellac and borax. Roll on the surface, and then brush over to even the coating. Dry. Next, paint on

three coats of Kremnitz white and pergamentine (water glass and glycerine). Steam at 248° F., and then calender.

John Mason's paper from plants and weeds. One of the most quoted individuals in the world of handmade paper, a lively and stimulating person whose *Paper Making As an Artistic Craft* has set vast numbers of individuals off on the road to doing-it-themselves, is John Mason. Here is but one of his approaches for consideration:

1. Chop local plants or weeds into

Ben Franklin's Approach À La Chinois

*I*n June, 1788, in a communication read before the American Philosophical Society titled "Description of the Process to be Observed in Making Large Sheets of Paper in the Chinese Manner, with One Smooth Surface," one Dr. B. Franklin wrote:

"In Europe to have a large surface of paper connected together and smooth on one side, the following operations are performed:

1. A number of small sheets are to be made separately.

2. These are to be couched, one by one, between blankets.

3. When a heap is formed it must be put under a strong press, to force out the water.

4. Then the blankets are to be taken away, one by one, and the sheets hung up to dry.

5. When dry they are to be again pressed, or if to be sized, they must be dipped into size made of warm water, in which glue and alum are dissolved.

6. They must then be pressed again to force out the superfluous size.

7. They must then be hung up a second time to dry, which if the air happens to be damp requires some days.

8. They must then be taken down, laid together, and again pressed.

9. They must be pasted together at their edges.

10. The whole must be glazed by labour, with a flint.

"In China, if they would make sheets, suppose of

four and an half ells long and one and an half ell wide, they have two large vats, each five ells long and two ells wide, made of brick, lined with a plaster that holds water. In these the stuff is mixed ready to work.

"Between these vats is built a kiln or stove, with two inclining sides; each side something larger than the sheet of paper; they are covered with a fine stucco that takes a polish, and are so contrived as to be well heated by a small fire circulating in the walls.

"The mould is made with thin but deep sides, that it may be both light and stiff: it is suspended at each end with cords that pass over pullies fastened to the ceiling, their ends connected with a counterpoise nearly equal the weight of the mould.

"Two men, one at each end of the mould, lifting it out of the water by the help of the counterpoise, turn it and apply it with the stuff for the sheet, to the smooth surface of the stove, against which they press it, to force out great part of the water through the wires. The heat of the wall soon evaporates the rest, and a boy takes off the dried sheet by rolling it up. The side next the stove receives the even polish of the stucco, and is thereby better fitted to receive the impression of the fine prints. If a degree of sizing is required, a decoction of rice is mixed with the stuff in the vat.

"Thus the great sheet is obtained, smooth and sized, and a number of the European operations saved.

"As the stove has two polished sides, and there are two vats, the same operation is at the same time performed by two other men at the other vat; and one fire serves."

Winifred Lutz's Approach

Some of that alchemy mentioned on page 143 is visible to the discerning eye in works Winifred Lutz transmutes from seemingly pedestrian materials.

Lutz is currently working in translucent paper to create unique forms cloaked in other-worldly mystery. She shared her experience about these new approaches:

"I have found that pulp made from fibers, such as sisal and milkweed silk, is best adapted to the forming of the very thin sheets necessary for real translucency. Sometimes I add a bit of bleached muslin or linen to these pulps, but never more than a third part by dry volume. (By the way, I never use cotton linters or half stuff, but make all my own pulp, since I have found I get better variety and truer control that way.)

"In terms of translucent paper, if made using modern techniques, I have two technical details that I've developed while working with thin sheets. First, it's normally quite difficult to lift the sheets from the blanket even when they have been treated with a good bit of pressure. Since drying the sheet on the blanket is time-consuming and does not always result in a desirable surface, I now separate the sheet by placing the blanket paper side down on a wet sheet of glass, rolling the back of the blanket with a brayer, and then lifting the blanket to leave the thin, wet sheet of paper adhered to the glass. This method results in the paper drying quite flat because it sticks to the glass. It also results in the paper having one very smooth, calendered-looking surface. It is easy to peel the sheet from the glass once the paper is dry. However, since it is necessary to use pulp that has been hydrated a long time for translucent sheets, if the paper is very thin (thinner than commercial wrapping tissue, for example), it tends to have a "memory" and to roll up in a scroll when peeled from the glass. This can be remedied by unrolling it between two slightly dampened blotters and pressing until dry.

"As to the second technical detail, a variety of effects can be achieved within the surface of a translucent sheet by freehand watermarking. Best results seem to come from using a spray gun on different settings and drawing on the wet sheet before it is couched on to the blanket. Adding inks or dyes to the water in the gun imparts a delicate edge of color to the watermark.

"I am using the term watermarking literally to describe the marks which I make with jets of water or water mixed with sumi ink. A surprisingly fine line can be achieved by this method, and it has a distinct advantage over the traditional wire pattern if flexibility and variation rather than uniform repetition are desired. However, it requires some practice with the water gun to create fine lines, since great speed in manipulation of the jet of water is essential."

...ch lengths and place in a ...e pan.

...Wearing rubber gloves, fill the sauce pan with water, and add, roughly, about 2 tablespoons (English dessert spoons) of caustic soda per quart of water.

3. Bring to a boil, and then allow to simmer for as long as is required to make the flora soft and pulpy.

4. Wash out all impurities from the pulpy mass, first through a coarse ball-type sieve and then through a fine sieve.

5. Beat the pulp in a mortar and pestle, run it through a meat chopper, and then place it in a kitchen blender or a Hollander (in small quantities) until your pulp is ready (as proved by testing).

6. Add the pulp to your vat, which contains an appropriate amount of water for the kind of sheets you wish to form, and dip your mould and deckle to make individual sheets of handmade paper.

7. Follow the same procedures as have been previously demonstrated to press, dry, size and/or calender your sheets. Or, let them remain as waterleaf sheets of natural color, unbleached examples of handmade paper from growing things.

7.
PROBLEMS
AND SOLUTIONS

The pulp sucked hard at the great mould, to drag it to its depths, but the man's strength brought it steadily forth; and then he made his 'stroke'— a complicated gesture, which levelled and settled the pulp on the mould and let the liquid escape through the gauze. Kellock gave a little jog to the right and to the left and ended with an indescribable, subtle, quivering movement which completed the task. It was the work of two seconds, and in his case a beautiful accomplishment full of grace and charm.

—E. Phillpotts, p. 18

Again, there is little agreement among papermakers (on a world basis through time and space) with regard to defects in handmade sheets of paper. In certain countries at certain times, what was regarded as one of many defects, such as a papermaker's tear—a drop of water from the vatman's arm or hand that falls on a newly formed sheet making the paper thinner and more transparent in that place and creates a veritable "crater" that disturbed past manufacturers of handmade paper—is, in some circles today, deliberately brought about for its esthetic qualities.

"In all old paper, as well as in the handmade papers of the present day, there is a considerable variation in the thickness and finish, and in single books the leaves vary noticeably in weight.

"The tone of the old paper was never entirely uniform, and, owing to the absence of chemicals in the manufacture, the grades of paper differed strikingly in colour." (Hunter, *History & Technology*, p. 224.)

On the other hand, with absolute justification, we probably are or should be concerned with flaws in our work (if we believe they are flaws); if we believe in the permanence of paper; if we give credence to our skills in multiplying perfect sheets to infinity; if we presume to produce papers with whatever properties are demanded of us; if we take pride in our craftsmanship and our capability to control the making of paper and not in being controlled by factors that, seemingly, may have been beyond the control of papermakers in the past. If, for example, we are faced with a heavy iron content in our water supply, if in preparing to beat our stock or in adding dyes or fillers during beating our pH factor is 5 or below (indicative of an unacceptable acid content), we are in a position to correct any or all of these deficiencies and others. Or, we should be.

It is apparent, therefore, that there are users who demand and receive perfect, constant, beautiful papers that do what they expect beautiful papers to do. And, we have individuals, also involved in the process of papermaking, who seek out the "accident," search for the unusual, deliberately depart from tradition for esthetic considerations, use paper as a medium in its own right, and attempt to push beyond the very outer limits—if there are any—of the medium.

I salute both groups—for their similarities and their differences—and believe that papermaking can serve their several goals.

	PROBLEM	SOLUTION
AIR BELL	This may be a blister, a foam mark, or foam on the surface, caused by poor lifting or forming at the vat or imperfect drying felts.	Change your felts, and practice, practice, practice lifting at the vat to form a better sheet.
BACKMARK	A stain or slight ridge that may have been caused when a still-moist sheet of paper was hung over a rack or rope (not of horsehair).	Use any one of a number of ways of drying paper, including that of air-drying single sheets on racks or shelves of fabric, or in spurs of four or so, or between blotters and changing the position and the number from time to time.
BAGGY	As in a man's trousers at the knee: a sheet with a pushed-in center.	Repulp and reform the sheet.
BELLS—see AIR BELL		
BLACKENING	Dark area or areas on sheets caused by too much or uneven pressure when sheets are calendered, or resulting from excess moisture in the paper while it passes through the calender rolls.	Adjust pressure on rollers; allow paper to dry further before calendering; repulp the sheets.
BLANKET HAIRS	When couching on new or unbrushed felts or blankets, stray fibers may catch on the sheet and remain on the surface. If not noticed by the artist, these hairs could become troublesome in that they could absorb watercolor or liquid ink and spread or fan out along the hairline.	It is good procedure when removing handmade paper from felts after the first pressing to brush each blanket or felt as soon as you remove the sheet. Further, it would be worthwhile to treat all felts to a proper washing with soap and water regularly.

	PROBLEM	**SOLUTION**
BLEEDING	Color edges dissolved by a liquid or staining out of a paper.	Check directions on dyes used; try more mordant; use a different brand of dye; experiment.
BLISTER—See AIR BELL		
BLOW—See AIR BELL		
BLUE SPOTS	Spots on the paper surface (of any color) caused either by the reaction of the particular dye being used plus rosin or by a poorly prepared dye, which may lead to a speckled effect, as in a brook trout.	Repulp the sheets; look at the "defect" positively—you may want to obtain that particular effect again. Change dyes, and try again.
BRISTLE MARKS	Marks caused by a stiff sizing brush when surface-sizing a sheet.	Use a softer brush and alter consistency of the size; try a less viscous sizing.
BROKE—See CASSIE or CASSE		
BRUISE	This may be identified as a dark area surrounded by wrinkles. It may be caused, among myriad reasons, by uneven pressure. It would be especially noticeable on the papers nearest the outside felt or blanket.	There is not much you can do with such a bruise, except to make sure the bookbinder's press, standing press, hydraulic press, or whatever other device you are using to remove water from the sheets initially is level.
BUBBLES	This condition may be brought about when air is trapped between the felt and the sheet. A negative force is brought about, unless the felts are all dampened.	Be certain, when preparing for couching, that all felts are dampened—not dry and not sopping wet.
BURST OR SMALL HOLE	As you might guess, a small hole could probably result from couching a too-wet sheet too quickly and unevenly. Or, it could also be the result of a wrinkled felt.	Straighten felts; couch carefully.
CASSIE, CASSE, OR BROKE	There is considerable disagreement among individuals as to differences between retree, outsides, and broke, or cassie. Obviously, some of the terms derive from the French; they all imply that there is a fault or more in the sheet of paper; that they are seconds and therefore sold under the regular price or used as the outside sheets or wrappings on reams shipped from point to point. Usually, the best of the lot are retree, the next outsides and the worst broke.	Many mills repulp these seconds and ignore the old ways and terms.

PROBLEM		SOLUTION
COCKLED AND CURLED	This undesirable wavy line quality in sheets of paper is the direct result of allowing a post of paper to remain in the screw-press (long ago) too long (among other reasons). This causes the edges of the sheets to dry while their centers are still wet. Papermakers have been and will be plagued with curled and cockled papers for generations upon generations. Paper breathes. It is hygroscopic. It cannot but be affected by humidity—always.	An old printmaker's suggestion would be to place these unhappy looking sheets between wet felts, press, keep overnight, and then transfer to a stack of clean blotters or equivalent, changing the blotters and the order of blotters until properly dry.
DEAD BEATEN	An overbeaten stock made into paper; it may be brittle and weak.	Keep better records of beating times and proportions of materials used; make tests more frequently.
DYE SPOTS	Uneven spots of strong-colored dye; may have been caused by undissolved dye or a fungus growth, which attracts the dye. May be a desired property of your paper.	Make certain all dyes are totally dissolved before using; keep all equipment and materials clean.
FEATHERED	A term used occasionally to describe a highly thinned-out deckle edge. Caused by pulp escaping under the deckle during the shake, or vatman's stroke.	This can be adjusted either by holding the mould more tightly to the deckle, sanding and planing the deckle or mould to create a tighter fit, or by following Henry Morris' directions on page 68.
FOAM DURING BEATING	One of many causes for the appearance of foam in the manufacture of paper may be an excessive alkalinity of the stock furnished to the beater. You may discover that you are using rag that was not fully washed of bleach, if you are using rag; or, that your recycled paper stock has not been properly cleaned.	One cure is to treat the stock in the beater with alum so as to reach a pH factor of 5.5.
FUR	In the process of couching, if done improperly (the sheets not couched one directly over the other in forming a post), fur or patches of pulp on the felt from the previous sheet will stick to the newly formed sheet.	The gentle art of couching takes much time and infinite patience. Try again. And again.
IRON SPECKS	Iron rust or specks appear on the surface of your sheet.	The culprit may be your water pipes, vat, or tank or any other piece of equipment and/or tool used in your process. Clean. Waterproof. May also be caused by an oversupply of iron in your water supply. In that instance, you, indeed, have difficulties. Consult local chemists.

	PROBLEM	**SOLUTION**
KNOT OR LUMP	Clots or tangles of very long fibers that create patches or dense shapes in the sheet when examined under strong light.	"The formation, or the lay of fibers in the sheet, can be improved by the use of certain gums that act as dispersing agents. For example, if the pulp fibers are a little too long, they will form clots or tangled lumps in the suspension, and the sheet made from this will be lumpy and cloudy. If you were to add about 1 to 2 % of deactylated Karya gum to the pulp, the clots would disappear, and the sheet would become quite uniform. Locust bean gum can be used, but it is not quite as effective. The paper chemical supply companies have recently come out with some synthetic materials that will act as well as the Karya gum. I think one might explain some of this with the analogy of trying to tie a knot in a bunch of eels. They are too slimy—the knots would not stay. The gums used in pulp seem to be surfactives that make the fibers slimy." (Harold H. Heller)
LINT	Unbrushed felts or blankets may deposit lint on the surface of new sheets. The lint may be picked off carefully with the use of a high-powered magnifying glass and a tweezers or left alone, if you are afraid of damaging a sheet.	Wash the felts regularly and brush them after each sheet is removed from a felt.
POLE MARK—See BACK MARK		
RIDGE	A high area in the paper surface.	Probably brought about because the vatman's shake or stroke was not brought to bear upon the pulp in the mould and deckle during the formation of the sheet. Kiss off the pulp and try again.
RING MARKS	Cloudlike effects of color in dyed papers; caused by bubbles in lifting or forming the sheet on the mould.	Prior to forming sheets in a vat of colored stock, be certain to check the manner in which you dip your mould. The process, when properly carried out, does not permit bubbles to form.
SIZE STAIN	Amber-colored or other specks caused by improper sizing.	Be certain, especially when dyeing stock in the beater, that the size or rosin (if that is what you are using) is fully mixed in with the stock.
STICK MARK—See BACKMARK		
WILD	A term employed by papermakers to describe a highly irregular, as opposed to a well-formed or well-closed, formation of interlocked fibers in a hand or machine-made sheet.	See Knot or lump.

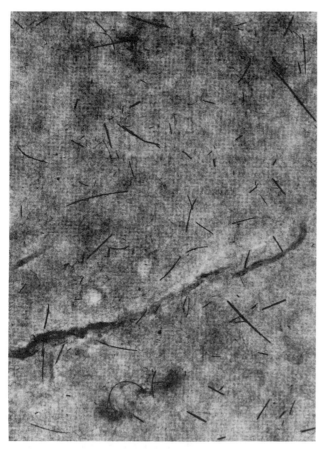

Bruised; poor couching; sheet detached.

Poor couching; sheet stuck to screen.

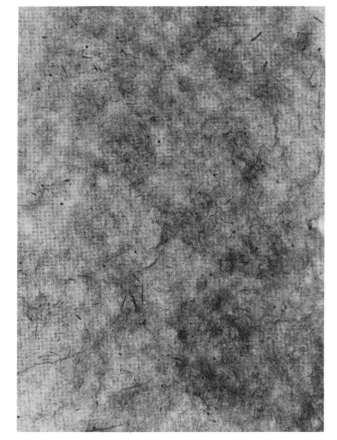

Poor formation; uneven.

A wild sheet.

ENVIRONMENTAL PROBLEMS AND SOLUTIONS

Before becoming involved in relationships between the environment and its effects upon paper, it would be useful to examine the role that man plays with handmade paper after it is formed and before and while it is transmuted into the carrier of visual ideas, or conceptual phenomena in works of art on single sheets or in books. (I am indebted to Francis W. Dolloff and Roy L. Perkinson's booklet, *How to Care for Works of Art on Paper.* Boston: Museum of Fine Arts, 1971, for this information.)

PEOPLE DAMAGE TO PAPER

Men and women cause problems with works of art on paper and with books for lack of using common sense in handling this fragile material. When lifting and moving a sheet of paper from one place to another, use both hands at opposite corners of a sheet to keep from bending, creasing, or tearing it. Make certain your hands are clean before you handle paper—if they are ink-stained and you are pulling a print, for example, use paper grips (folded 2 × 4 inch paper strips) in each hand to protect the sheet from being permanently stained. Never stack unmatted pictures on top of each other, but separate them with acid-free tissue or glassine. Store matted works in solander boxes or wooden blueprint cabinets. Given the choice, do not ship unmatted works on paper in a roll when you can ship or carry them flat between heavy chip boards or 1/8 inch Masonite panels. For fine works of art on paper, never employ any adhesive but methyl cellulose when an adhesive is required for matting a drawing or print.

I wish to thank Don Farnsworth for his note which stated, "Mix 1/2 pound of methyl cellulose to 5 gallons of water. This paste has an indefinite shelf life, is insect-proof, pH neutral, impervious to fungal attack."

NATURE DAMAGE TO PAPER

Insects, rodents, lizards, fungi, the very light that emanates from the sun or from manmade light substitutes, temperature, and humidity, the smog-infested air we and paper breathe are some of the factors considered in this section below.

Insects. Cockroaches, silverfish, termites, and woodworms are the prime enemies of paper. Their appetites are enormous, and, I guess, papermakers provide delectable meals for these voracious critters.

Cockroaches, my dictionary informs me, are members of "the family *Blattidae*, . . . many species of which are troublesome pests in houses . . . especially, in warm climates. They are usually nocturnal in habits; they can run very quickly; some long-winged species can sustain flight."

You will find cockroaches in warm, dark, dank situations. They are particularly fond of painting mediums composed, in part, of sugar, paper, leather, glues, and fabrics to name just a few items in their diet. They particularly damage the surfaces of these materials and are nuisances.

Good household cleaning is essential, especially the basements and attics, if you want to avoid these insects; if discovered, powdered insecticides should prevail.

Silverfish, and again I consult my dictionary, are "a small, wingless, silvery insect of the genus *Lepisma* . . . found about houses and sometimes injurious to sized papers, starched clothes, etc."

Silverfish also prefer dank, dark, warm situations; they, too, move so rapidly that they are difficult, if not impossible, to follow or find before they have done considerable damage to your favorite works on paper. These silvery rascals will eat their way right through the surface of your prints and drawings to devour glue and starch sizes they contain.

Termites and woodworms describe "any of the numerous, pale-colored, soft-bodied, social insects of the order *Isoptera*—white ants. . . . Some of the species of southern and tropical Africa build great nests of clay twenty feet or more in height. . . . They are often very destructive to buildings, books, etc. . . ."

These little monsters can eat their way through your entire home through their network of tunnels that undermine any structure, devouring not only the wood framework and furniture within, but also anything made of cellulose—and, unfortunately, that includes paper.

Temperature and humidity. Ideally, and there is no such condition, it would be nice if we lived in a world where the temperature is moderate all year long and the humidity rarely, if ever, got above 40%. Unfortunately, there are few, if any, such places, and when the humidity rises above 70%, mold sets in. Now, if you are fortunate enough to live in a moist climate, you can alter the internal climate through the judicious use of air conditioners and dehumidifiers. And, everyone can, from time to time, open windows to air certain rooms that greet you with a musty smell when first you enter.

Foxing. The term we use to identify mold growth in papermaking is termed foxing. These lifeless, rusty spots on papers are usually caused by the chemical reaction between the colorless iron salts present in most papers and mold growth. The higher and more prolonged the humidity, the greater the degree of mold or foxing. Mold will nourish itself with paper fibers and sizing to the point where it will reduce the strength of the sheet.

Ideal water, if there is such a substance, contains no calcium, less than .05% iron, and is soft. Given the proper temperature and humidity conditions plus an iron content in the water of .05%, foxing will occur in your paper. Given an

John S. Copley (1738-1815). Study for George IV when Prince of Wales. *M. and M. Karolik Collection. Reproduced from* How to Care for Works of Art on Paper *by F. W. Dolloff and R. L. Perkinson, Museum of Fine Arts, Boston, 1977. This is a very good example of paper fading—note the darker areas, where the frame originally covered the edges.*

unlimited budget, you can have a deionization device installed to rid your water supply of unwanted iron content. Or, if you don't make great amounts of paper at a given session, it is not inconceivable that you could purchase distilled water for that session. On a limited budget, however, we must do the best we can with the tools and materials at hand—at least, our tools and equipment can be rust-free and should be so maintained.

Treatment for foxing entails instant removal of the sheet to an arid environment, exposure to the sun and the air for an hour or so to allow free circulation, or enveloping the sheet in an airtight container with some crystals of thymol (a fungicide) for several days to kill the mold.

High temperatures, in and of themselves, deteriorate paper. Thus, at the risk of being overly repetitive, it is suggested you make paper in that ideal place where humidity and temperature remain constant.

The sun and manmade light. One of the elements we take for granted and about which we, apparently, know very little, is light—especially as it affects paper and works on paper.

Without fear of contradiction, I can state that light fades all works on paper (prints, watercolors, drawings, etc.) and also fades or discolors the paper itself. It does not matter at all whether we are talking about sunlight, fluorescent light, or incandescent light of any degree. Dim, strong, weak, powerful, hazy, brilliant—light fades and discolors papers and works contained on them. Visible light rays damage paper and pigments, but the shorter ultraviolet rays cause the most severe damage.

The most dramatic example of this problem I witnessed was caused not by exposure to direct sunlight, but by indirect morning light reflected within a particular room. The right and bottom of a fine blue paper were covered by

the original mat, thus protecting these areas from fading—the rest of the color was completely gone.

Ideally, works of art on paper or virgin sheets should be viewed under no more light than would be required to read a book in visual comfort.

When paper is curing, after manufacture, it should not be placed in direct sunlight, nor should it be kept in the same place indefinitely. Move papers from time to time, as you would pictures in your home. Both will profit from having been placed in different locations for periods of time. Ultraviolet light in all these instances is the enemy of paper—respect the enemy and take the proper precautions. Knowing that fluorescent light radiates ultraviolet light, be certain such lights are covered with plastic sleeves. If you have a stack of paper curing, place a sheet of plexiglass on top of the stack, since it does not allow ultraviolet light to pass through.

Smog and other air pollutants. Urban papermakers, beware the sulphur dioxide in the atmosphere, aside from the other smog pollutants. Sulphur dioxide can be absorbed into your paper and be converted to sulphuric acid, which will not evaporate even when removed from the source of the pollutant. Severe brown stains, brittleness, and decomposition will follow in the wake of sulphur dioxide gas on your paper.

Short of moving to an idyllic retreat to make paper as have the Kollers (HMP), the Clarks (Twinrocker, Inc.), and the Wilfers (Upper U.S. Paper Mill), air conditioning is your best defense. Also, through tests of your papers by local chemists in your area, you may decide to put alkalies in your pulp to offset the acid danger.

SPECIAL PAPER PROBLEMS RELATED TO PRINTING

There is or should be a one-to-one relationship between printing and the particular paper selected for a

specific printing process. The intimacy between the graphic arts and papermaking demands that certain of these problems be further described at this juncture.

Curl or cockling. As mentioned above, no one, thus far, has solved the problem of curling and cockling paper. Since paper is manufactured from cellulose fibers, since it is hygroscopic, since humidity affects the living, breathing sheet of paper by swelling or contracting the fibers that form it, the challenge confronts all papermakers. Each one has his own solution, which, it appears, is successful solely under the conditions that exist in that specific place and no other. Marvelous!

Paper curl is more obvious and a more difficult problem to solve when it concerns lighter-weight papers than with heavier stock.

See the previous pages for suggestions that may lead to success, and, at the very least, give you something to talk about if failure leads you to therapy.

Hickies. These unplanned-for ugly spots may appear in printed or nonprinted areas and may be due to faults in the paper manufacture, the printing, or a combination of the two.

When examined under a high-power glass, if the hickie has a dark center surrounded by an unprinted area it may be attributable to a printing problem. A tiny particle of dried ink (not our problem) can affix itself to an ink roller or an offset blanket and then be transferred to the paper in the course of printing.

However, if the center of the hickie, when examined under magnification, reveals itself to be a *white* center—it is our problem, a paper problem, in which paper fibers pulled from the sheet's surface, lint, or dust have affixed themselves to the surface of our formerly pristine sheet. Especially in offset printing, this is a problem. The fibers are water-loving, and they sit on top of our paper surface; thus, they accept water, repel ink, and give the printer a headache. He can usually clear it up by reducing the tackiness of his ink or the speed of the press.

Piling. Piling results from a deficiency in the bonding within the internal fibrous structure of the paper. It may come from a lack of moisture resistance in the paper coating. It may, in other cases, be caused by a certain paper's adhesive that appears to come loose and then transfers to the offset blanket of an offset lithographic press.

You can readily see what would happen when the additives (starch, fillers, etc.) transfer to and start a build-up on the nonprinting areas of the offset blanket. Obviously, the quality of the subsequent printed areas suffers enormously. It would be visually evident in blurred halftone dots, mottled printing, and a severe loss of the image.

Scumming. When ink adheres to the nonprinting areas of the lithographic stone and a master lithographer is at work, we must assume that the problem is inherent in the pH value of the paper. It must be so low (so acid) that the paper counter-etches the stone or plate and causes scumming. Curiously, many of the well-known mould-made papers are slightly acidic in nature.

Tinting. This is almost the opposite of scumming—and it, too, occurs in lithography. If the pH of the paper is excessively alkaline, it may neutralize the normal, slightly acidic condition of the dampening water, or the so-called fountain solution. The net result of this unhappy combination of paper with slightly acidic water on a stone or plate causes a uniform tint to appear all over the noninked areas and the ink pigment to bleed.

INTERNAL PROBLEMS AND SOLUTIONS

With proper care and feeding, we know that paper (in the Western world) can last for at least 500 years or so. A visit to any major Print Room anywhere will confirm the fact that you can examine and be awed by fifteenth century prints pulled on still-fresh-looking papers. (Some papers from China are at least 2,000 years old!)

We are also aware, I believe, that a great many paperback books, trade books, newspapers, newsletters, journals, and so on become, in time, not newsprint but some brittle substance that once was paper, but now is untouchable and may not be read or handled for fear it will disintegrate. This fragile yet tough substance may be reacted upon by internal and external conditions, some of which we may be able to control.

Since it would require the proverbial six-foot shelf of books to enumerate all of the problems and enemies of paper, the following examples may prove useful: Dard Hunter (*History and Technology,* pp. 225–227) mentions short tears, knots, water drops, hairs from felts embedded in the paper, drops of size, bubbles, wrinkles, small holes, blurred watermarks, ruffled laid lines, and pieces of rust from some part of the vat. . . .

If a term you are looking for is not listed below, see the Glossary.

8.
EXPERIMENTAL APPROACHES— A GALLERY

Cade. ". . . I am the besom that must sweep the court clean of such filth as thou art. Thou hast most traitorously corrupted the youth of the realm in erecting a grammar school: and whereas, before, our forefathers had no other books but the score and the tally, thou hast caused printing to be used; and, contrary to the king, his crown, and dignity, thou hast built a paper-mill. It will be proved to thy face that thou hast men about thee that usually talk of a noun and a verb, and such abominable words as no Christian ear can endure to hear. . ."

—Shakespeare, *King Henry VI Second Part,* Act IV, Scene VII

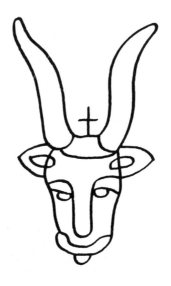

Inca Hand No. 2, 1974.
Cast paper and watercolor.
22 x 24 inches.
Collection Dr. Roger Guillemin,
La Jolla, California.

This gallery of experimental paper-making information, in visual form and the printed word, is designed primarily to reveal the use of hand-made paper as a medium in its own right.

There are, as we have already noted, numerous problems inherent in making sheets of paper. On the other hand, in a world where technology plays perhaps a too significant role in our lives, it seems refreshing (at least to me) that more and more contemporary artists have become attracted to handmade paper either as a meaningful support for their two dimensional visual ideas or as a most flexible, three-dimensional, self-controlled, self-manipulated, permanent medium of expression.

Recent exhibitions of handmade paper prints and unique paper works in prestigious museums and galleries the world over strongly suggest that establishment institutions are beginning to offer the new-old medium their official stamp of approval. Colleges and universities in greater and greater numbers believe they are slightly out-of-date if papermaking is not taught as a respectable studio discipline by at least one member of their staff. There is a flurry of activity in any village, town, or city when a papermaking workshop is announced.

As it would be impossible to illustrate the paper works of all of the contemporary practitioners in the field, the following selection is based upon my private biases, which I readily acknowledge.

GARY H. BROWN

For a number of years, Gary H. Brown (b. 1941) has explored and attempted to synthesize disparate phenomena related to the hand: the hand as regarded by the basic principles and concepts of western perspective, including the Golden Mean; all of the possible meanings of the visual, plastic, psychological, physiological, and esthetic considerations of the hand from many cultures; televised images of the Apollo and Soyuz handshake; and perhaps Michelangelo's fresco of that moving, creative, pointed finger.

The presence of the paperwork shown strongly suggests, to my eyes, that craft and process (certain critics to the contrary) are intrinsic qualities of what we call art.

Summer Sierra Fault, *1977. Colored handmade paper. 32 x 34 inches.*

JOHN BABCOCK

If you have ever wondered about the surface quality of unknown, strange planets in our galaxy (including the planet we inhabit), that sense of wonder can be curiously satisfied on examination of this recent work by John Babcock.

Babcock's two, and occasionally, three-layered portraits of faults and fissures in the crust of Spaceship Earth cannot but remind you of micro San Andreas Faults, especially if you are aware of and sensitive to the topography of California.

Since the artist does most of his work outside his California studio, except in very cold weather, I strongly suggest the essence of the young, rugged geological environment of the West has seeped into the pores of his skin and permeated his mind, to be reborn as statements of magic about his world.

He has worked out methods and formulas, working procedures and tools, equipment and supplies that meet his needs uniquely and inexpensively. (See pages 63 and 98.)

SAM GILLIAM

When a painter of the stature of Sam Gilliam (b. 1933) decides to explore the medium of paper in Joe Wilfer's Upper U.S. Paper Mill, he does so with all stops pulled, as his recent *Great American Quilt Series* verifies.

Winner of many awards for his works, including a Guggenheim Memorial Foundation, Gilliam is a widely exhibited, well-documented artist. The energy in this work virtually vibrates—Gilliam explodes, and the viewer in tune with the work finds echoes of the sounds of one hand clapping.

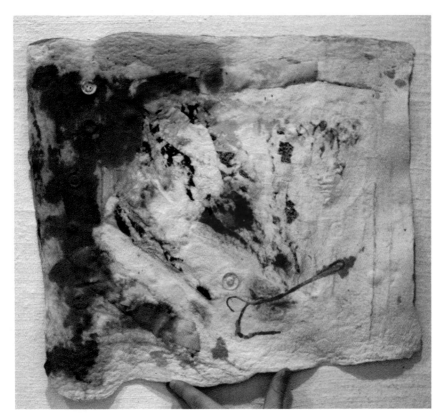

Great American Quilt Series (1 of 40), 1976. Paper embedments, color, handmade paper. 18 x 24 inches.

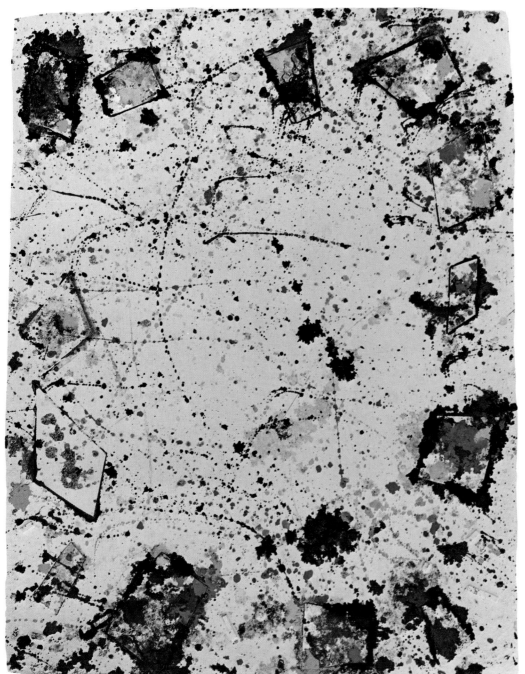

SAM FRANCIS

Even a brief look at this work assures me that Sam Francis (b. 1923) has had a love affair with hand-made paper. As in the past, he masterly focuses our attention on the perimeter of the handmade sheet with lush color forms, leaving the center as a negative force or set of tracks that leads the eye, again and again, to the margins.

Untitled, 1976. Handmade paper, watercolor, and oil.
22½ x 20 inches. Collection Ann and Garner Tullis.

Colored Paper Images—XII,
Colored pulp laminated to handmade paper.
46½ x 32½ inches. Published and printed
by Tyler Graphics Ltd. Photo Steve Sloman.

ELLSWORTH KELLY

A fully realized collaboration between the artist Ellsworth Kelly (b. 1923), the master-printer-publisher Ken Tyler, and artist-master paper-maker John Koller, resulted in a series of colored pulp laminations of extraordinary quality.

RICHARD ROYCE

Richard Royce is filled with technical information, ideas for prints and paperworks yet unrealized, words spilling out faster than my hand can record them. They are all related to two-dimensional and three-dimensional cast paper prints, felting super-size, unique, air-dried sheets of paper, collages, casting techniques, rubbings and/or castings of "slices of life," such as the front windows of his house. He also concentrates on parts of various monuments embossing variations, using air-brushes from different angles with complementary colors to achieve a particular sought-after effect.

The intriguing technical achievement in this paper work is the manner in which Royce solved the inking of the inside of the curved surface. After he worked the inside plaster surface with dental tools and all manner of gouges to achieve his esthetic goal, he inked a rubber balloon which was then used to transfer the ink to the curved sphere.

Sea Sphere. *Cast paper.*
18 inches in diameter.
Courtesy the artist.

116

JOHN KOLLER

One of the few individuals I have been privileged to meet in this dollar-oriented world whose life, art, and work express an enviable unity is John Koller (b. 1943). He and his wife, Kathleen, are the proprietors of HMP, one of the more successful, ongoing paper mills in the United States—a beehive of effervescent life, hard work, creative adventure, instant gourmet food, and stimulating talk.

Koller lives and works in a home surrounded by his own forest. For those whose cynicism leads them to believe that the individual no longer can survive today without loss of integrity, dignity, and such, I suggest the Kollers prove otherwise.

Yellow and Grey with Maroon Paint,1976. Dye, water color, and dry pigment on cotton. 24½ x 33 inches. Courtesy the artist.

JUAN MANUEL DE LA ROSA

In between trips that range from Mexico City to Paris and the rest of the continent, through many cities in the United States and elsewhere on this spinning globe, Juan Manuel De La Rosa (b. 1945) somehow manages to produce works on and in paper that are bold, dynamic, colorful statements derived from his life experience.

In *Tied to Time,* De La Rosa reveals the latent power in Mexico and in Mexican art. No matter what the medium or the form, it appears to me that a long tradition of making and forming is a most potent phenomenon with regard to the young artists of that country.

"This work has been realized using the same material used by the ancient Mayas and Aztecs—*amatl* paper (the inner bark of a moraceous tree)—plus rag and cotton pulp, which gave me a very strong and resistant paper. I thus had the opportunity of using it not only as a plane surface but also as a deep area—I used different elements *inside* the paper, such as a small old clock, cords, etc. As a matter of fact, this work gave me the feeling of enormous potential leading to future magical discoveries with handmade paper."

Tied to Time, 1976. Dye, collage, embossed amatl, rag, and cotton handmade paper. 24 x 19 inches.

HAROLD PERSICO PARIS

In a world where artists and others occasionally come together to reinforce brotherhood, soul-cleansing, ear-cleaning, eye-washing, and sundry limited or long-range goals, Harold Persico Paris (b. 1925) has, seemingly, gone it alone.

Coincident with his 1975 exhibition at Smith-Andersen Gallery in Palo Alto, the artist wrote, "I do not see my art as estranged from my heredity and from previous times. I have sought to visualize the paradox, in graphic form, of myself as the man and the separate entity, at times, of myself the artist. The pattern of my life is one of constant seeking for the elusive—often impossible to describe—romantic, visionary, poetic, closed personal door. Making paper, coloring the pulp, imbedding the images within the paper, the casting, the forming into the paper molds of the color and the objects was a ritual which not only immersed the object, but portions of myself."

In a letter to me in July, 1977, Harold P. Paris closed with the following statement: "I don't know if you know about this, but in 1951 and 1952 I worked with Douglass Howell, who introduced me to handmade paper which was subsequently used in my *Hosannah*. Since 1975, I have completed about 150 pieces in paper."

The Hidden Screen for Raindrops, 1975.
Handmade paper.
12⅜ x 10 inches.
Courtesy Smith-Andersen Gallery, Palo Alto, California.

KENNETH NOLAND

Kenneth Noland (b. 1924), whose work is represented in major museum collections from Los Angeles to London and elsewhere, evidently found that working in colored paper pulp provided him with still another medium with which to solve self-imposed problems.

Untitled, 1976.
Cast paper pulp.
34½ x 34½ inches.
Collection Ann and Garner Tullis.

MICHAEL PONCE DE LEON

Michael Ponce de Leon (b. 1922) has been working with handmade paper made expressly for his collage intaglios for many years by master-papermaker, Douglass Howell. These four feet square, *one-inch thick* sheets of pure linen are more than just an ordinary ground for his work—they are intimately related to the total concept of each collage intaglio worked by the artist. To prepare this precious paper for printing one of his multicolor works on his self-designed hydraulic press, the artist adds a tablespoonful of plaster of Paris and a teaspoonful of pure gelatine to about two quarts of water. All of this liquid is applied to the paper with sponges for dampening, just before placing it on the matrix and submitting it to the forces of the hydraulic press.

Echo 7859402.
Collage intaglio.
19⅛ x 18¾.

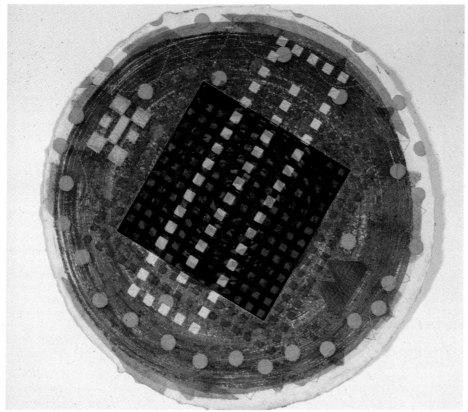

ALAN SHIELDS

Although the name Alan Shields (b. 1944) and his works have been known to me for many years, I have never met the artist. This man has been an important force in the unorganized, unknown, quiet revolution that has affected both printmaker and papermaker.

Since you cannot discuss one without the other in Shield's case, perhaps we should consider the two activities as one here.

He has woven his prints with handmade papers, embossed lithographs, and tried innumerable approaches (and still does) to that which he feels he must make with his hands. He was closely associated with the Shenanigan Press, the logo of the Jones Road Print Shop and Stable, as well as the Upper U.S. Paper Mill.

Shields' Shield. *Colored handmade paper, mixed media. 30 x 12 inches.*

FRANK STELLA

Working with publisher Ken Tyler and papermaker, John Koller of HMP, Frank Stella (b. 1936) moved from his shaped canvases and similarly approached two-dimensional color lithographs to test the water, so to speak, and pushed beyond the surface plane, using carefully considered hues, values, and intensities of dyes on hand-made paper.

JIM PERNOTTO

In collaboration with Joe and Mike Wilfer, Jim Pernotto (b. 1950) opens the world of funky color paper works to titillate the eye and challenge the mind. Pernotto initially carves his work in clay, makes plaster casts, and then "pulls" his paper pieces from the casts. "They are then painted with dyes, acrylics, colored pencils, rhoplex, and further embellished with marbles, sequins, etc. The work is then a cotton piece with the durability of canvas, and it takes color like the finest watercolor paper."

That's Our Gary, *1977. (Above) Painted cast paper. 72 x 82 x 9 inches. Courtesy the artist.*

Paper Relief—Kozangrodek, *1975. Dyed and collaged handmade paper. 26 x 21½ inches. Printed and published by Tyler Graphics Ltd.*

The Lyric Song Series, Lester and Billy, *1976. Dyed, multicouched, handmade paper. 23½ x 18½ inches,*

CHARLES R. STRONG

"I think of this as painting with pulp. The whole process of painting is contained within the papermaking process. No paint is added at a later time."

Thus did Charles R. Strong (b. 1938) end his letter. It appears that he, indeed, works with as many as 6 to 10 different colored pulps, couches them one upon the other until he has obtained the visual goal he was seeking, and then presses the work in the standard manner.

NEDA AL-HILALI

One of the most involved, tortuous, yet visually clear, modes of working with paper is practiced by Neda Al-Hilali (b. 1938). Al-Hilali subjects the papers she works to chalk, charcoal, dyes, inks, paints, pencil, rhoplex, and varnishes. At present, she sews, carves, crops, and interweaves these wild fragments of what once were pristine sheets by taming them with "weathering, kiln-baking, branding, sanding, and compression through steam-heating, sledge hammering, and pressing with steel rollers.

"The results do not refer back to the material, but rather radiate away from it in numerous directions, alluding to, implying, connoting, or suggesting an expanding range." The very large work of paper shown here truly defies description; it has a presence of its own and throws out challenges, on many levels, to its viewers.

Untitled, *1977. Mixed media. 12½ x 8 feet. Courtesy the Allrich Gallery, San Francisco. Photo Blair Paltridge.*

SUZANNE ANKER

It is dangerous to read into works that which may be merely in the viewer's mind; yet it is tempting to suggest that Suzanne Anker (b. 1946) delights in posing forces one against the other—ebb and flow, a certain easy rhythm in counterpoint with a "difficult" movement, the seemingly natural against man-made paradox and existential truth. But, I digress. Anker's cast paper piece speaks eloquently for itself.

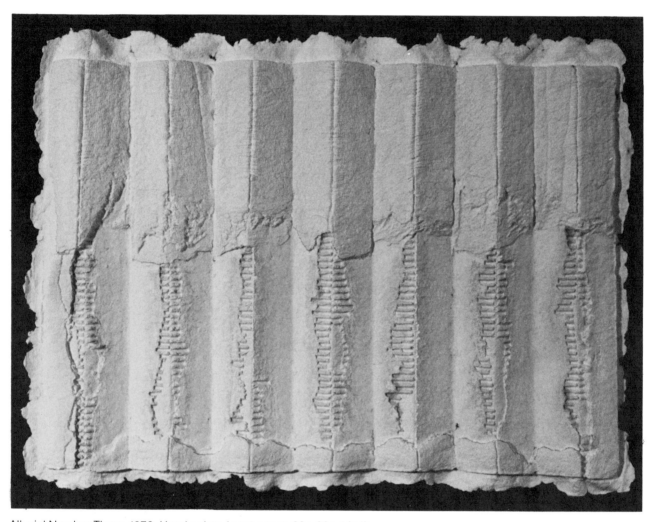

Alluvial Number Three, *1976. Hand-colored cast paper. 22 x 30 x 1 inch.*

DIANA ARCADIPONE

Papermaker for the last five years, Diana Arcadipone enjoys experimentation. This work was made by embedding alfalfa seeds between two couched sheets. The seeds burst forth (in the dark) during the drying process.

"After germination, I exposed the paper to the sun so the sprouts would grow tiny green leaves. The leaves actually grew out of the paper surface. Later, they turned brown and dried, but still created an interesting image."

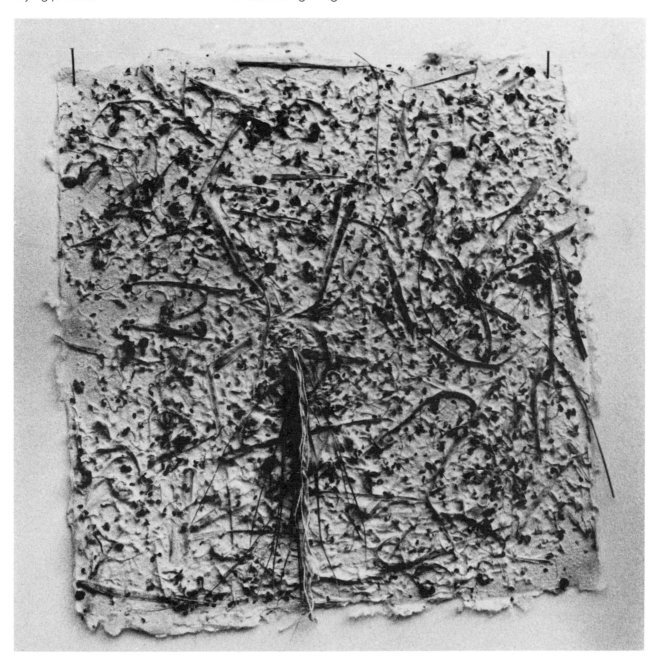

Paper Sprouted Paper Sprouted Paper Sprouted, *1977. Handmade rag paper. 12 x 12 inches.*

CATHERINE BABINE

In describing her working methods, Catherine Babine (b. 1949) writes that her collages are "made by double and triple couching, alternating layers of paper and collage elements. Textures are formed by the use of different sizes of screens and by moulding with various instruments after the paper is couched. Also by the depth/lack of, of the various collage elements. A favorite, though not necessarily wise, technique of mine is the use of wire, nails, screens, washers—anything containing iron—for their invariably stunning reaction with the paper and water. Somehow, I do not expect these collages to deteriorate. My oldest piece using iron has not changed in over a year."

Babine, in other paper works, uses variations of a technique seen in a film: "Dyed pulp is released from a vessel with a small hole, using a finger to control the flow, on to a large screen. White pulp is then couched from a smaller screen over the pulp that has been poured. Finally, the large screen is couched, leaving the image face-up and reversed."

Manuscript Series No. 4, 1977. *Handmade paper, collage. 14 x 19 inches.*

LAURENCE BARKER

In writing from his studio in Spain, Laurence Barker touched upon several areas of interest to all papermakers, especially on the troublesome question of linters:

"Linters fit the bill perfectly for those who don't want to have to beat their heads against the wall because they have their plates to proof, their moulds to cast. But I don't view it as an either/or situation. I go along—perhaps fatuously so—thinking there is time enough to do the two jobs very well, paying full attention to papermaking. It's fast enough work you know: in under three hours on my machine, I can tickle to death 20 pounds of raw material that will provide me with sufficient pulp for over 100 sheets of Imperial or more. Papermaking is fast, clean work, and there should still be plenty of time to be an artist.

"My collages are some of my most recent works, and they might be considered to be the antithesis —or the obverse—of the casting technique, inasmuch as the work is done face-up and is collage in nature. I make shaped paper, print woodcuts, linoleum cuts, or etchings on these shaped pieces and embed them on fresh, wet background sheets. Once dry there may be, and often is, pencil and ink work."

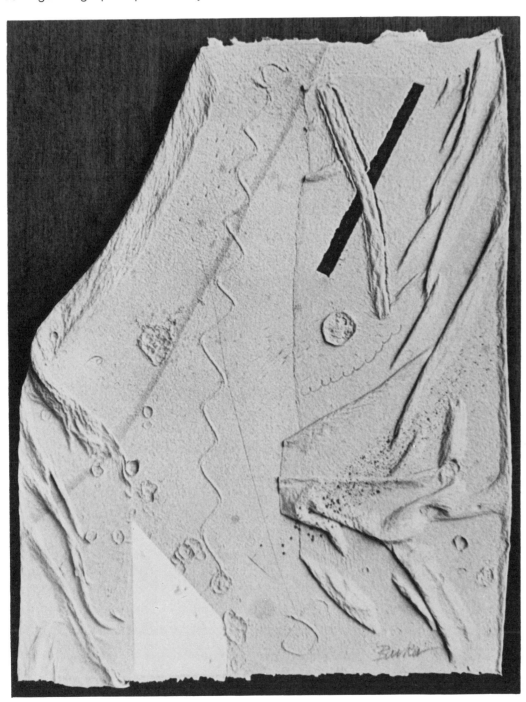

Untitled, 1977.
Handmade paper
collage.
23 x 29 inches.

KATI CASIDA

Kati Casida, in collaboration with Don Farnsworth, has produced various series and individual pieces having references to sounds—from whispers to the roar of the surf. She has also created paper works which require strong yet delicate paper, combining various printmaking, drawing, and other mediums within their two and three-dimensions. She folds her collages "after the first pressing, when the paper is like damp felt." Only when you have worked with paper lovingly and with dedication do you arrive at such formulations.

Ribboned Sea, *1975. Handmade papers, silkscreen, and woodcut-printed ribbons. 28½ x 43½ inches.*

VIJA CELMINS

Untitled, *1975. Lithograph on handmade Twinrocker paper, 17 x 20 inches. Courtesy, Cirrus, Los Angeles.*

Wrapped Venus, *1975. Etching on handmade paper. 28 x 22 inches. Courtesy, Landfall Press, Chicago.*

WILLIAM B. CLARK

William B. Clark (b. 1948) has been quietly making paper in the southwestern desert for a number of years. His works are reminiscent of totemic forms, enigmas cloaked in mysteries.

"The use of paper in my work has led me through a variety of media. First, paper served as a surface to paint on; then it took me through the various print media; and now to a multidimensional product in and of itself. In each instance, paper was the element of exploration. Paper has emerged through this exploration as a spontaneous and plastic medium allowing collage, assemblage, and manipulation. The finished two or three-dimensional totalities yield rich color, tone, and textural elegance. There is something magical about manipulating paper that restores that innocent creativity I remember as a child."

Paper Art, 1977. *Handmade paper, graphite, and jute. 28 x 20 inches.*

RONALD DAVIS

The work of native Californian Ronald Davis (b. 1937) is not unknown in New York's Museum of Modern Art, London's Tate Gallery, the Albright-Knox Gallery in Buffalo, and other major museums.

His distinguished exhibition record is peppered with works seen at the Venice Biennale, Documenta in Kassel, Germany, and places here and abroad.

The work shown here, a multicolored intaglio (14 colors) done on HMP handmade paper by John Koller, plays freely with minimal form in space; it successfully weds bitten line, burred line, and grainy tonal areas with scalpellike precision in a liberated, many-hued, color-pulp environment.

Bent Beam, *1975. Color intaglio on multicolored paper. 20 x 24 inches. Printed and published by Tyler Graphics Ltd.*

130

DOMINIC L. DI MARE

He is in love with the ritual of making one sheet of paper at a time in the traditional manner, yet something occurs in the process that revolutionizes the rags, flowers, plants, and feathers found on his property in California and transforms them into magic lures for the mind and eye. Dominic L. Di Mare (b. 1932) says it all in his own words:

"I began making and using handmade papers in 1970. My attraction was not that I could make papers on which I could draw or write—but that the process triggered feelings and memories that seemed central to my being. The making of paper allows me to examine and record these memories and feelings in a way that brings them into focus. Much of my youth was spent with my father, a commercial fisherman, fishing off the coasts of California and Mexico. The whole process of my making paper mirrors that important part of my life. Living aboard a boat and the catching of fish—the struggle that is so much a part of that existence seems to be reflected in the magic of the reaching into the water, 'catching' the pulp, and ending with the frozen wave."

Bundle, Twelve Earth Meadow Poems— Marchello, *1977. Handformed rag and artichoke blossom paper, hawthorn wood, raffia, raccoon bones, yellow-shafted flicker feathers, and handmade felt (New Zealand fleece). 9 x 10 x 4½ inches (and a 20 inch "tail"). List Collection, New York.*

TOM FENDER

The bridge between fiber art and the art of handmade paper is tenuous. In the work of Tom Fender (b. 1946), especially in a recent series titled *Packages,* the bridge is crossed; the areas are integrated to create mysterious, age-old, bundles of handmade paper chips wrapped (or are they trapped?) in equally enigmatic fibrous tendrils.

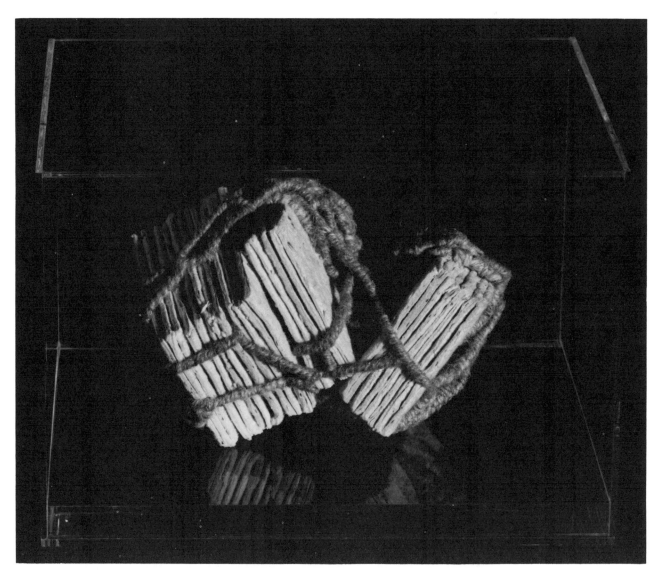

Leo Package, 1977. Rice paper, mulberry paper, and wool. 6½ x 4 x 5 inches. Courtesy The Allrich Gallery, San Francisco. Photo Jan Stedman.

NANCY GENN

Painter, sculptor, and artist who works in handmade paper, Nancy Genn is proud owner of her own Hollander (a Valley beater), a new hydraulic press, and all the tools, materials, and such with which to make paper in her own studio.

"My work in paper dates back to an earlier time when I was doing large, mural-size paintings on paper—oversize rag paper I imported from England. These were successful and were followed by a period of doing lithographs in the 60s both in New York and in Paris. My drawings often became assemblages.

"Having enjoyed working with paper, I progressed naturally to working directly in freshly formed paper as a medium. I have had the fortunate experience of developing some of the techniques of paper works with Garner Tullis at Santa Cruz, with John Koller in Connecticut, and particularly with Donald Farnsworth at his Oakland paper mill. In 1975, I acquired my own Hollander beater and have been doing a good deal of my paper works at my own studio."

Now that she has her own papermaking equipment, it will be interesting to guess at the effect it will have on her work, since she will be able to macerate rags even in the middle of the night, if she chooses. Is there a hint in Genn's statement?

"Paper is, for me, both a two and a three-dimensional material. A unique contribution to the medium that I have been developing over the past three years is linear tear-ups of varying widths in freshly formed paper. In some cases, the paper torn away will reveal a colored layer underneath or maybe a negative space. There is a sense of mystery to the way the other layers are revealed.

"The layering of the material, an important quality in the development of my work, is accentuated when it is embossed. I also used the embossed line to emphasize a shape or a natural edge.

"I often use paper pulp as pigment to give the emotive result of painting by grinding the pulp finely. For a large-scale colored area, though, the cotton linter is dyed and separately prepared in my Hollander beater. In this case, I make a point of not finely grinding each individual color to avoid a flat surface and to allow the mixture of the fibers to give vitality to the material. This gives a rich natural surface.

"The texture and stiffness of the material that makes it so sculptural influences the way in which I have developed the form. In combining these personal developments, my aim has been to create a clear visual statement that is possible only in paper, using it in a new and original manner. Paper is an old material, but here I have used it in a totally different way, happily bringing together my experience as painter and sculptor."

Naka. *Drawing on handmade paper laminated with color, threads, and rice paper. Courtesy Susan Caldwell Gallery, New York. Photo Jeremiah O. Bragstad.*

PETER GENTENAAR AND PATRICIA GENTENAAR-TORLEY

Peter Gentenaar of Rijswijk, Holland (born 1946), writes, "My interest in paper derived from printmaking—a heavily embossed etching gave me ripped prints. So, I started to look into the paper. I wanted to make it thicker and accomplished that by means of vacuum-forming my sheets and layering them immediately after they came off the screen. I do this layering either in a mould or on another surface, and press the paper dry.

"If it's formed in a mould, I will take the form out after it is dry enough, let it dry thoroughly, and possibly sand or polish the surface. If paper is formed on any surface and is pressed dry as well, I'll treat it like wood: saw it, sand it, plane it, etc., and build sculptures out of it, combining it with other materials.

"The layered paper has my special interest. It brings me the feeling of layers of the earth, an old feeling that is very natural as a material to work with—it dictates

(Left) Transformer House, 1977. Colored, couched paper, sawed, lacquered, and assembled on a metal frame 5¼ x 3⅓ feet (center square 6 inches thick.) Photo Jan Schurman.

(Right) Big Molleton, 1976. Moulded, handmade paper with molleton surface, 11⅞ x 8⅛ feet. Photo by Jan Schurman.

the forms you're going to make from it.

"The excursion my wife (b. 1949 —she's Californian from Mountain View) and I took into feltmaking ended with the cotton-surfaced paper reliefs, which are very soft-looking, structural, and colorful. All the different approaches to working with paper I have found so far I still use, and I keep finding new varieties."

The works of Peter Gentenaar and his wife, from my vantage point, appear to go far beyond technical virtuosity into the mysterious area of art. Though, without Gentenaar's ability to see beyond the accident of the ripped sheets of paper; without that certain sense that allows an artist to seize upon a negative phenomenon to make of it something new, exciting, and positive, there would merely be another individual who simultaneously happened upon vacuum-forming as one of many experimental approaches.

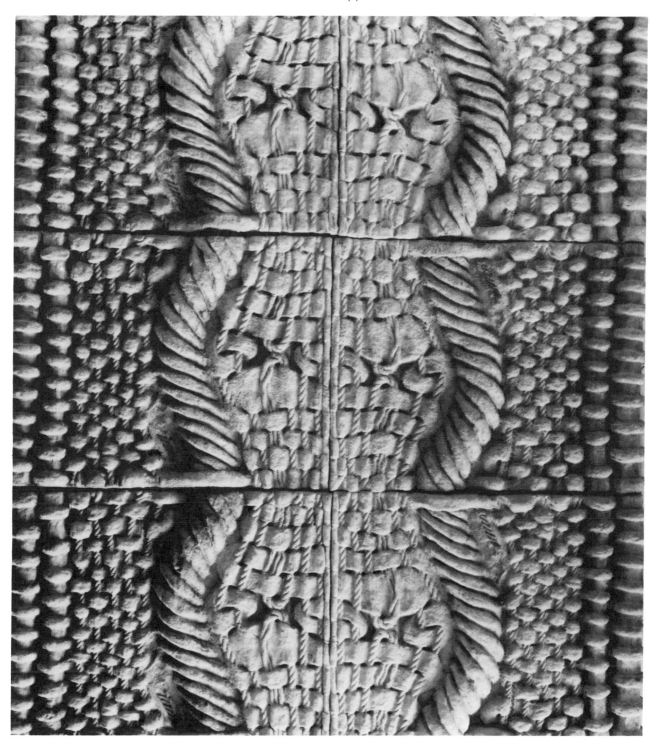

CAROLINE GREENWALD

Caroline Greenwald truly loves to work in, on, and with paper—as her progress as an artist reveals. She has a delightful sense of humor, is a prodigious worker, and speaks with enthusiasm and grace about that which she loves.

"Caroline Greenwald's work with handmade paper began with her printmaking on to both surfaces of imported oriental papers and suspending these translucent works into space. She then began to use these papers alone, no longer willing to impose an image on to these unique, responsive materials.

"Her earliest works of papers were eight foot translucent panels into which she embedded lines and then suspended these panels to create an environment through which the observer could move. While the scale of Greenwald's work is becoming more intimate, the viewer is still involved with the art object, as with a series of books which unfold.

"Greenwald's artwork utilizes the natural colors of the translucent oriental papers, shadows created by embedding fibers into the work and manipulating the original sheet into layers of long narrow folds in combination with opaque areas of rag pulp and her own handmade papers."

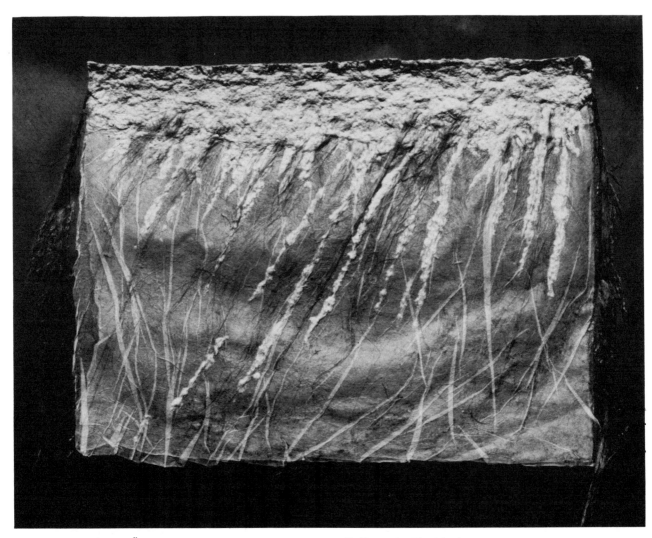

California Edge Series #4, 1977. Translucent oriental paper, raw silk fiber. 14 x 17 x 1 inch.

CHARLES N. HILGER

Charles N. Hilger (b. 1938) is quite specific about mediums and messages, and, in the work shown here, we can observe the precision and clarity of his ideas within the limits he sets for himself: "The whole of contemporary society uses paper as a carrier of statement and image. By reversing that condition, my image is the paper that becomes the statement. I create my own material from pure white cotton and work with that material alone. To involve color and/or foreign materials would negate my statement."

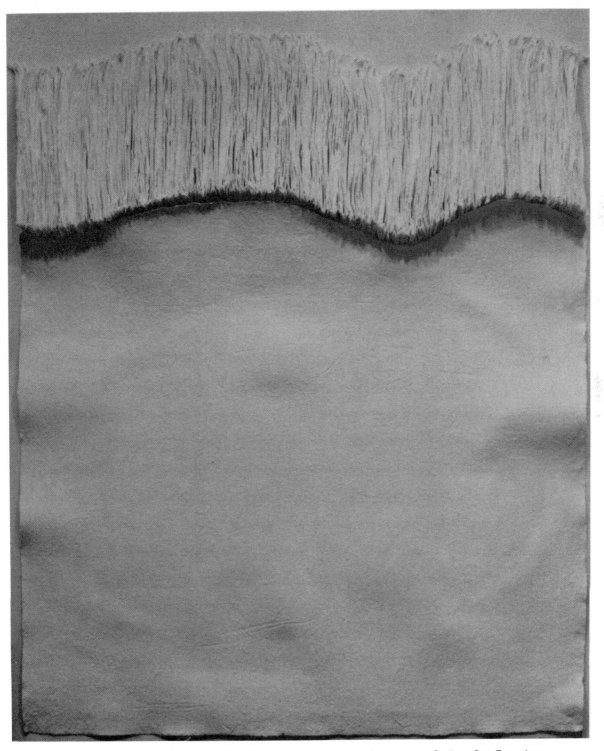

Starting Down, *1977. Handmade paper. 70 x 58 inches. Courtesy Smith Andersen Gallery, San Francisco.*

CLINTON HILL

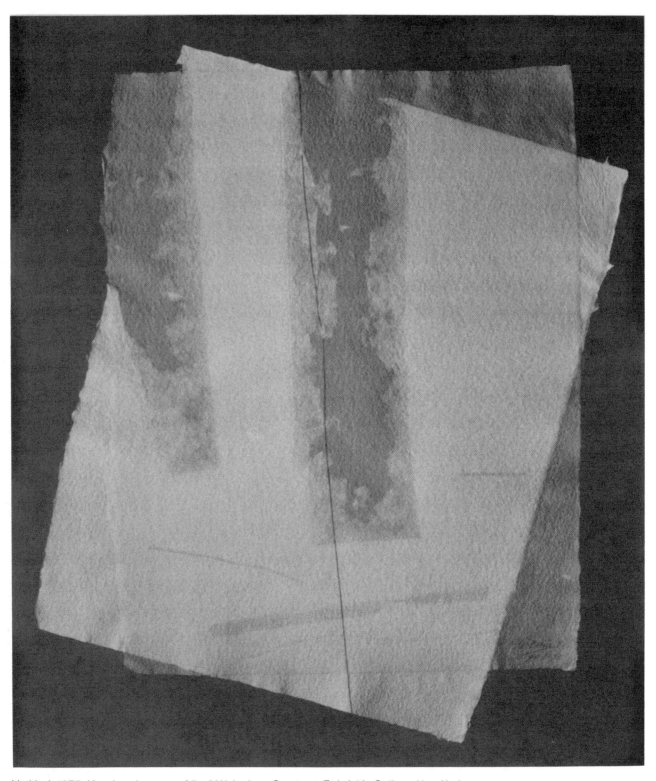

Untitled, *1975. Handmade paper. 25 x 22½ inches. Courtesy, Zabriskie Gallery, New York.*

SANDY KINNEE

Sandy Kinnee (b. 1947) traveled the long road from printmaking to painting back to printmaking and papermaking before he found the proper materials to fit his visual ideas.

His recent works include editions of multicolored, shaped handmade papers incorporating etching or screenprinting plus carborundum, glass beads, silver leaf, and various other materials.

"About three years ago, I found my paintings were getting much too large (8 × 10 feet) and fragile. This was especially a problem for exhibiting the work. To make them, the paint was poured on to room-size release surfaces, built-up, dried, and then a backing or support was attached.

"But just because my paint had dictated configuration—my ink did not have to do the same. I discarded cutouts, plastics, plaster, and decided to make my own shaped paper with as inconsistent a deckle as possible. All my paper is shaped. No square or rectangular material."

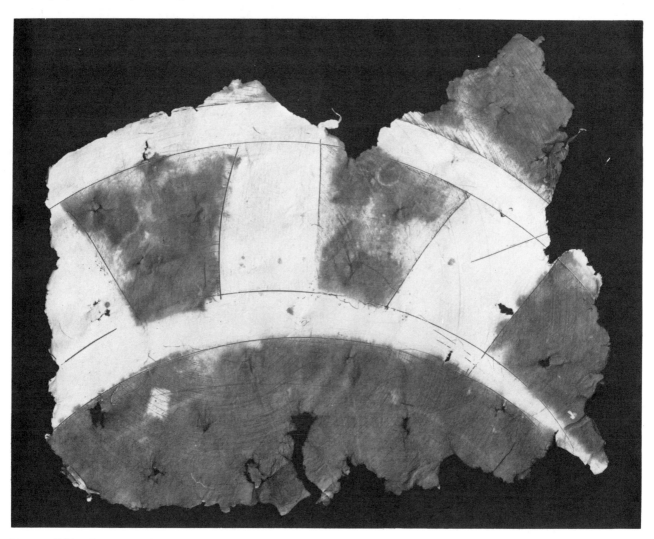

Monet's Bridge Fan, 1977. Colored etching on handmade, hand-dyed paper, ed. 28. 24 x 19½ inches. Photo Roger Alken.

KRASNO

While wandering through the multitude of Parisian rare book and print shops in 1967, my eye chanced upon a castpaper print that entranced me. Yet, I walked on. Hours later, I followed my feet back to the print and book display to find the work, housed in its plastic box, still unsold. Obviously, I bought it.

If memory serves, Krasno is a South American artist who was then living and working in Paris. Unfortunately, that is all I know about him.

What is most unusual about this work, an impossible-to-photograph paperwork, is the way light is caught, reflected, and bounced about *in* the relief forms, which are other handmade papers with holes or wafflelike grids affixed to the initial sheet, probably when it was still damp.

Untitled, *1967. Handmade paper, 18½ x 12¼ x 2 inches. Collection the author. Photo Merrylee Stephenson.*

LOUIS LIEBERMAN

Louis Lieberman (b. 1944) is a prolific, seemingly indefatigable artist whose "less-is-more" works grace the collections of major museums and private owners in Europe and the United States.

"*Untitled* is part of a series of cast paper reliefs that I have been working on since 1975. I consider the pieces to be paper sculptures rather than prints, as they are sometimes referred to. The works are done in small, limited editions of from three to eight pieces at most. They were originally in-tended to function somewhat as maquettes for my larger fiberglass wall reliefs, but have obviously become works which must be considered on their own.

"The pieces are made by spraying wet paper pulp (made by macerating etching paper in a kitchen blender with water) using a modified paint spray gun, into a plaster mold. The original forms are worked first up in clay and then into a plaster model from which the final paper casting mold is made."

Untitled, *1976. Cast paper relief, 14 x 18 x ⅝ inches. Collection Metropolitan Museum of Art. Photo courtesy the artist.*

KATHRYN McCARDLE LIPKE

Moving from fibers, printed textiles, and the dyeing of textiles more than prepared the way for Kathryn McCardle Lipke (b. 1939) to feel not unprepared to work intensively with Twinrocker, Inc. and, the following year (1975–76), with the "last remaining Swedish hand-made paper mills in Lessebo."

"My approach to paper pulp has been a feeling of going back to a more basic or primitive form of fiber, responding to the macerated threads themselves and building out—in a sense going within the fiber itself and then working out, building and meshing together. I am concerned with the tensions created by the thin layer of mesh. At the same time, I feel I am using fiber in all its forms—thread, cloth, pulp."

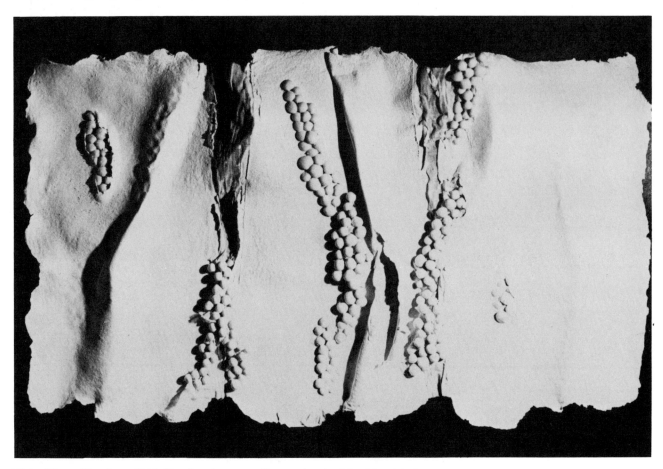

White Winter Morning, *1976. Handmade paper, fiber and pulp. 60 x 35 x 4 inches.*

WINIFRED LUTZ

Winifred Lutz (b. 1942) began making paper in 1957 "out of banana peels, pineapple tops, and seaweed (land and sea fibers). My interest was in papermaking as alchemy—converting dross into something of mystery. After I had refined my papermaking in graduate school with Laurence Barker and had spent some time using my paper as a vehicle for my prints, I became interested in the process for its own sake. Papermaking became the working counterpoint to my sculpture, both an extension and an alternative—Apollo and Dionysus, the paper being Dionysian. Worked with directly for itself, paper is compellingly sensuous and responsive and the making of it is an alchemical event." See page 101 for a technical description of this work.

First Reading, 1976. *Translucent watermarked sisal paper scroll on pinewood screen faced with gampi. 29½ x 21 x 3¼ inches.*

SØREN MADSEN

Working feverishly since the arrival of a Hollander beater early in 1977 at Toronto's Ontario College of Art, Søren Madsen (b. 1953) has accomplished, assisted in, collaborated upon, or published no less than 35 editions of works ranging from 25 to 150 handmade sheets each, in a period of three months.

Despite the furious pace at which he works, Madsen finds time and a certain zen-quality to suggest in linen and hemp that "life is *not* a straight line."

Life is Not a Straight Line, *1977. Handmade linen paper with hemp rope. 11 x 13 inches.*

JOANNE MATTERA

On a rigid grid that stands 15 perforations high by 17 wide (some mystical number if multiplied or manipulated arithmetically in some way?), Joanne Mattera (b. 1948) plays off an irreverently interposed, delicately counterpoised series of cotton threaded knots—to the visual delight of the onlooker.

When I quizzed Mattera about her work, she responded, "Sewing cotton, linen, and silk thread on to paper allows me to draw in a way which synthesizes my lifelong experience with fiber, on the one hand, and my traditional fine arts education on the other."

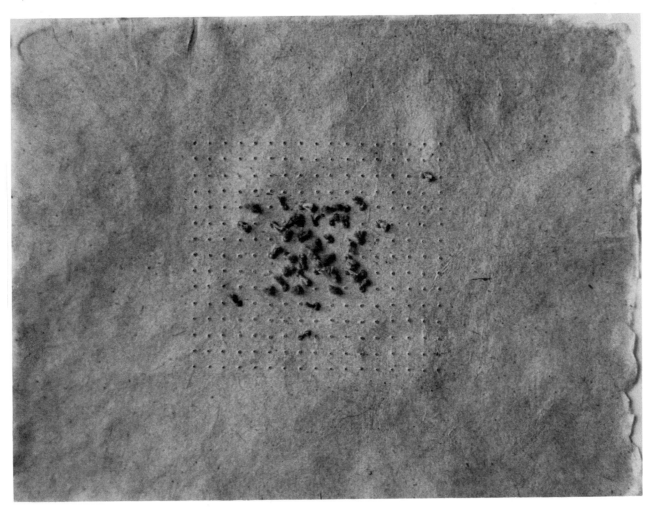

Drawing with Perforations and Knots, *1977. Handmade paper and cotton thread. 17½ x 20½ inches.*

ROBERT MOTHERWELL

In exploring collage, a medium often used by Robert Motherwell (b. 1915), feeling appears to be the primary basis for decisions with regards to cutting, shaping, arranging, and tearing. In the recent past, Motherwell searched out the essence of elements around him: what is the blueness of the Gauloise blue? The silkness of silk? The qualities of paper?

In the sugar lift shown, two carefully placed tusche strokes record the movement of the artist's hand over the stone, accompanied by the blobs and spatters that usually occur, or are made to order, when working with a brush filled with grease held in suspension. There it is.

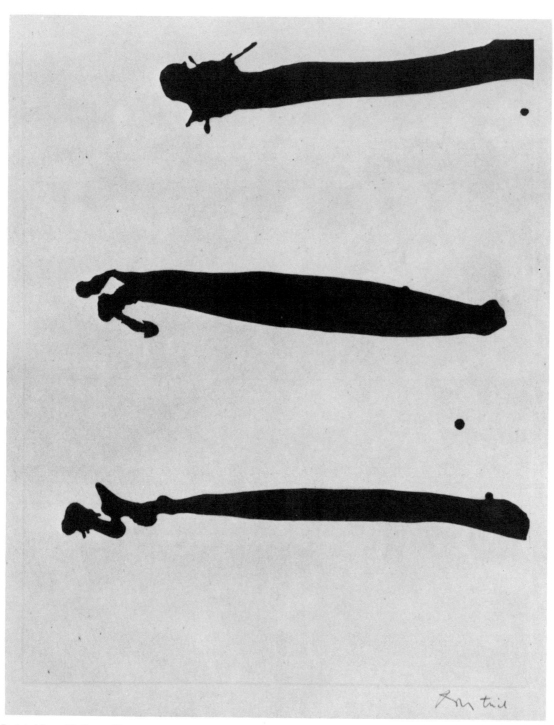

Dutch Linen III. *Sugarlift aquatint. 29 x 25½ inches. Courtesy Brooke Alexander, Inc., New York.*

LOUISE NEVELSON

I still carry traces within the "furniture of my mind" of the 1975 solo exhibition at Pace Gallery by Louise Nevelson (b. 1900). There were two floors of works: white sculptures were displayed in a pristine white environment on one level; on the other level, you entered a black space to find it filled with black sculpture. It was an experience that defied description, unless you wished to flirt with the impossible. There were also works *in* and *of* paper.

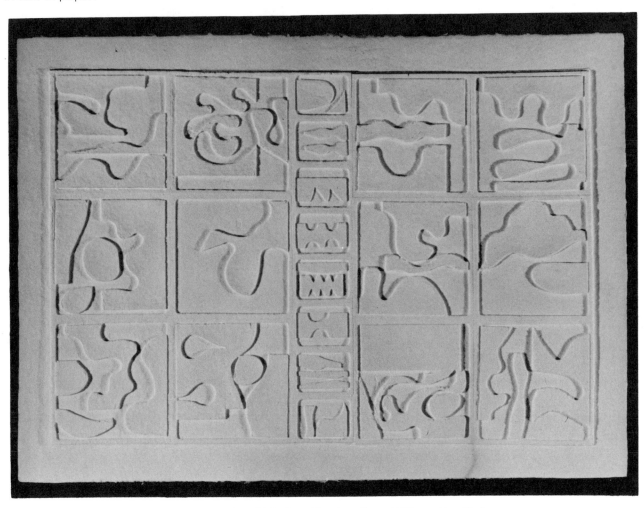

Dawn's Clouds, *1977. Cast paper relief. 27¾ x 39 inches. Courtesy Pace Editions, New York.*

ROBERT NUGENT

When I quizzed Robert L. Nugent (b. 1947) about his works in paper and why and what they were about, he replied, "I believe they are just about me, things I'm interested in, what I'm about—past and present. I've always been interested in antiquity and the way the earth harbours secrets from the past. I've always loved books—having books —more than just reading them. They, too, hold secrets. And I love to get mail (kept letters); as a small boy I used to write for every free catalog and brochure—anything that would come free of charge and just for the asking. I would rush home from school to my stack of mail. Later, that was translated into love letters and the need to write them. I know this is where the *Kept Letter Series* came from; I'm still deeply involved. The letters have now become captain's logs, mariner series. Still letters—notes— journals.

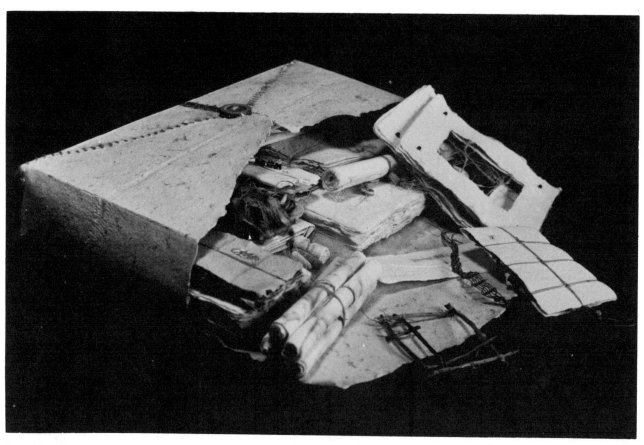

Ancient Mariner Series, Adamant, *1977. Handmade mulberry and rag papers, wood, handmade felt, raffia, sea urchin spines, ostrich egg shells, beads, buri root, conté, bone, and gold leaf, 13½ x 14½ x 3 inches.*

NANCE O'BANION

Nance O'Banion (b. 1949) takes us into her confidence with this statement:

"I have always enjoyed the idea of reconstruction and manipulation in my work: creating one form, destroying it or altering it to give life to another form. In many of my papers, this concept is manifested by actually constructing one paper, cutting it up and remaking it into another by casting it in more pulp. I'm very attracted to the dangerous aspect of this process, to the possibility that something may not work when it is reconstructed and that I may have destroyed a piece that was good in its former incarnation.

"Handmade paper by nature is transformed from a liquid state to a solid, a process which allows for a tremendous amount of freedom. By my approach to papermaking, I can place the raw, liquid pulp anywhere, on any surface, with any texture to create the texture/surface quality in the paper that I desire. Once the first stage is done and it is dry, I begin to manipulate the paper surface to develop the constructed layers and create images. I feel that this final metamorphosis gives life, energy, and emotion to the works. When the 'personality' is complete, I no longer feel a part of my work as it has its own identity and character.

"The notion of scale is also very important to me. It is my goal that every paper have a strong graphic power at a distance, as well as a seductive delicacy up close. Because of my technical approach to paper, I am able to make works that vary greatly in size: the smallest works are around 6 by 6 inches and the largest single pieces are 3 by 4 feet. The modular concept is also open to me, and this has resulted in pieces 8 by 8 feet. I decide on the size of the works according to the personality I am developing, or I often use the scale as an initial limitation and work spontaneously from it.

"Needless to say I have been completely seduced by handmade paper and have not yet felt I have exhausted its potential." It is hoped that O'Banion will continue to find working with handmade a never-ending adventure.

Disintegrating Structures, *1977. Handmade paper, 2½ x 3 feet. Courtesy the Allrich Gallery, San Francisco.*

CHARLES PACHTER/MARGARET ATWOOD

Canadian painter and printmaker, Charles Pachter, in collaboration with a poet of the same country, Margaret Atwood, created four superb, limited edition folios: *The Circle Game* (1964); *Kaleidoscopes: Baroque* (1965); *Talismans for Children* (1965); and *Speeches for Doctor Frankenstein* (1966). Pachter designed and printed six other folios from 1964 to 1968, some on handmade paper he produced at the Cranbrook Academy of Art while studying under Laurence Barker (1964–1966).

"The editions were limited to 15 or 20 copies, the paper being specially created as a matrix for the different print media—litho, screen, etching, and foundry type. The goal was to marry words and images in a visual bond so one without the other would seem deficient. The luxurious texture and resiliency of handmade paper provided a perfect environment for this, and led to several discoveries, the most noteworthy one being that the 'kiss' impression of hand-inked type from a flatbed letterpress on the nubby handmade paper surface was inimitable and—to this printmaker—immediately addic-tive. There was simply nothing quite like it. Running a sheet of handmade through a litho press compacted the fibers and gave a beautifully smooth silky finish in contrast to the puffy feel of freshly dried paper. The textural combinations provided a glorious setting for some intense and vivid poems out of which the images flowed naturally."

Pachter's work is alive and well in Toronto. His work grows and changes with the growth and development of one of the most stimulating cities in North America.

Pebbles From Expeditions, *1965. One of eight poems and lithographs on hand-made paper in a limited edition folio. 22 x 30 inches.*

Nude on Iron Bench, *1975. Etching on handmade paper. 22 x 28 inches. Courtesy Landfall Press, Chicago.*

ROLAND POSKA

The Fishy Whale Press, printmaking, and papermaking have long been identified with the bearlike proprietor-artist, Roland Poska (b. 1938). His paintings, lithographs, and papermaking activities and works are equally known to aficionados from the village of Carefree, Arizona to the largest provincial town of them all—New York City.

Poska wrote me that his primary interest is "my new series of prints and my book, *Where Forces Meet or Part* (a definition of the Deckle-Edge, Literal)." When you ponder that partially glimpsed statement about the literal definition of the deckle edge, Poska emerges as the many-faceted individual I have always believed he is.

He is direct, forthright, and, in his paper works, almost overpowering. The viewer standing in front of one of his *papestries* (a term he coined to define his paper tapestries) will feel totally engulfed in a part of Poska's world; one of his recent works is more than 100 feet wide, composed of 18 separate but related panels.

First Frost Reflections, *1975. Handmade paper, color papestry. 7 x 72 feet (18 panels).*

DEBRA E. RAPOPORT

In addition to her collaborative works and performance pieces, Debra E. Rapoport constantly surprises others with her myriad approaches to papermaking: from pristine and imaginative uses of purchased handmade as shown to seemingly wild, though well-controlled mixtures of ribbon, stitchery, pulp, vinyl, acrylics, seaweed, thread, handmade felt, and hardware.

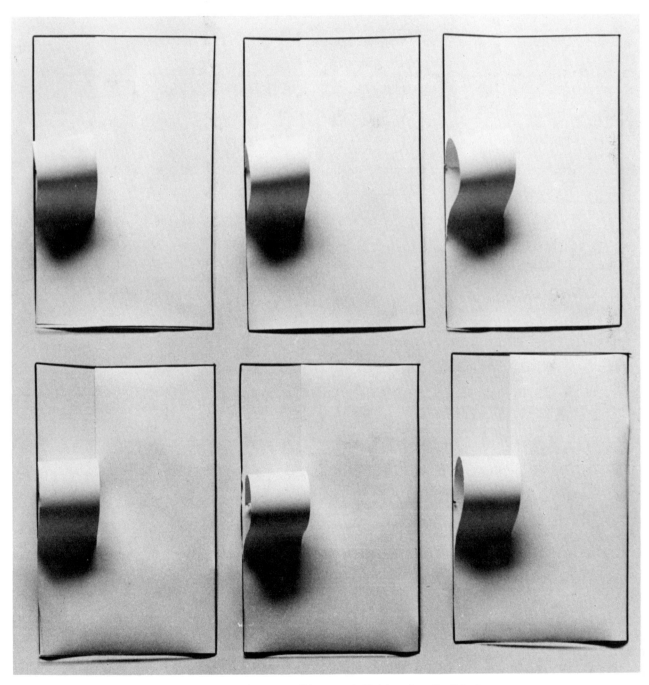

6 Unit Paper Piece, *1976. Purchased handmade paper, ribbon, and stitching. Area 8 x 8 feet.*

ROBERT RAUSCHENBERG

Considering the visit of Robert Rauschenberg (b. 1925) to the Richard de Bas paper mill in France, during which the series *Pages* and *Fuses* were realized in 1974 under the aegis of Ken Tyler (a name long associated with printmaking and now with paperworks), it is curious that all manner of artists soon make their way toward the nearest paper mill or university facility containing a Hollander beater or enroll in a professional workshop to learn the rudiments of working with this fascinating new-old material.

What motivated and still motivates so many human beings to explore new approaches to a two thousand year old medium? Is this a reaction against the technology of our time? A reflection of it? Does it express a yearning for the simpler things? A desire for a "hands on" approach once more?

Vale (State—Pages and Fuses). *Screen print on tissue-laminated pulp. 19½ x 24¾ inches. Courtesy Gemini G.E.L. ©, Los Angeles.*

BARBARA SHAWCROFT

A teacher from Atlanta, widely exhibited from Baton Rouge to Homer, Alaska, Barbara Shawcroft's works are collected by artists and curators. As a weaver, painter, ceramist, photographer, and papermaker, she has had her work seen and reviewed internationally.

Irony of ironies, when the Japanese view Shawcroft's paper works in an international exhibition both in the Kyoto Museum of Fine Arts and the Tokyo Museum of Fine Arts at the end of 1977.

Numero Otto, *1976. Handmade paper, mixed media. 30 x 12 inches.*

FRED SIEGENTHALER

Since we expect to find many approaches to papermaking, as is true in all of the visual and plastic arts, it should come as no surprise to discover that Fred Siegenthaler (b. 1935)—whose paper mill is situated in Rumisberg, Switzerland on the Southern pitch of the Jura Mountains—produces papers and paper works of his own invention.

He states, "I am proud to tell you that I am the artist who developed a new expression in modern art, the so-called incorporation of objects into paper, or dematerialization, or giving a soul to the enclosed object. Please note that the incorporated objects are totally covered with paper on all sides."

A papermaker since 1967, Siegenthaler has traveled to Nepal, Northern Thailand, the islands of Fiji and Hawaii, Mexico, and Ecuador to observe, firsthand, the handmade papermakers or the producers of tapa and other bark papers. All of these experiences influenced his work done at his present mill in Switzerland. In the work shown, note what occurs when iron objects are embedded in paper slurry and the rust stains move from the inner structure of the paper to the outside.

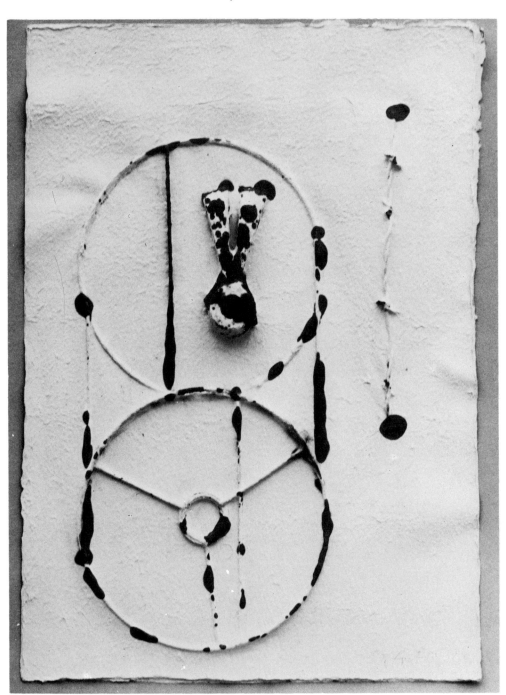

No Title, *1976.*
Iron objects and
barbed wire embedded
in handmade paper.
20 x 28 inches.
Photo Jacques Ludwig.

ANDREW SMITH

Andrew Smith needs no interpreter. "A detail of the world's first photographic watermark, depicting a young lady's face. This detail was taken from a circular piece of watermarked paper 5 inches in diameter (formed in tin cans without the use of any sophisticated papermaking equipment).

"The pulp was chemically produced from spruce bark. It is a facsimile of the original edition, which I produced in the late sixties from marijuana. Eventually the entire edition was smoked. The image I used was originally created for an advertisement to sell O Cannibus flags, which I devised upon immigrating to Canada.

"Its yellow color, shiny fibrils (clumps of fibers), and freeness—illustrated by its unroughened fibers and pinholes—differentiate it from white cotton paper. I exposed a very fine silk screen, coated with photographic emulsion, to a piece of litho film. Recorded on the film was a background image of rosettes formed by overlapping three halftone screens (80 lines per inch), at different angles, combined with a mezzotinted portrait. The washed-out areas of unexposed emulsion were filled in with lacquer, and the remaining emulsion stencil was washed out. This gave me a photo stencil insoluble in water. At the time it was made, there was no photo emulsion available that was impervious to water as there is now. However, pulp likes to stick to the new waterproof emulsion, upon which the sheet of paper must be formed.

"Though I created this idolatrous image, I find it immoral to chop down trees and enslave people to reproduce it in a book. Therefore, anyone should be duly warned from plagiarizing its technique, for fear of succumbing to the insanity that possessed me when I created it."

Mary Jane, *n.d. Photographic watermark, actual size. 2¾ x 4⅛ inches. Courtesy The Isaacs Gallery Ltd., Toronto.*

WILLIAM WEEGE

Founder and proprietor of The Jones Road Print Shop & Stable, William Weege, with the assistance of a patron, has put together the ultimate printmaking workshop in which all traditional and experimental mediums may be practiced by Weege and invited artists and stretched to the limits of their individual and collective imaginations.

It is not unusual to find a print emanating from the workshop that was made on or with handmade paper, used color litho, etching, screen printing, die-cuts, a bit of letterpress, photoscreenprint, vegetable printing, flocking, stitching, dye-dipping, hand coloring, and anything else you can think of.

Untitled *(detail), 1976. Collage embedments in handmade paper. 40 panels, each 24 x 18 inches.
Courtesy collection of Dr. Donald Eiler, Madison, Wisconsin.*

RUTH WEISBERG

When you pull an almost-impossible to control two-stone lithograph made of tusche washes on a chamois-colored, irregularly shaped piece of handmade paper produced by Twinrocker, the result, shown above by Ruth Weisberg (b. 1942) is both appealing and enigmatic. The cool and hot forms and colors weave in and out of the surface plane to play upon your senses. In addition to the regular edition of forty prints pulled on handmade paper, Weisberg informed me that three were printed on chamois skin.

Chamois, *1974. Color lithograph on handmade paper. 42 x 42½ inches.*

IRENE WHITTOME

Montreal artist Irene Whittome has "made sculpture with paper for the past four or five years."

Altar, photographed in progress, was one of a number of works based on *The White Museum,* a theme that embraces Whittome's works in recent years. Using paper and other materials with which to define and refine structuralist and emotionally charged approaches as well as to seek out unity rather than separateness, Whittome's work provokes the viewer to search deeply within, to probe one's internal computer.

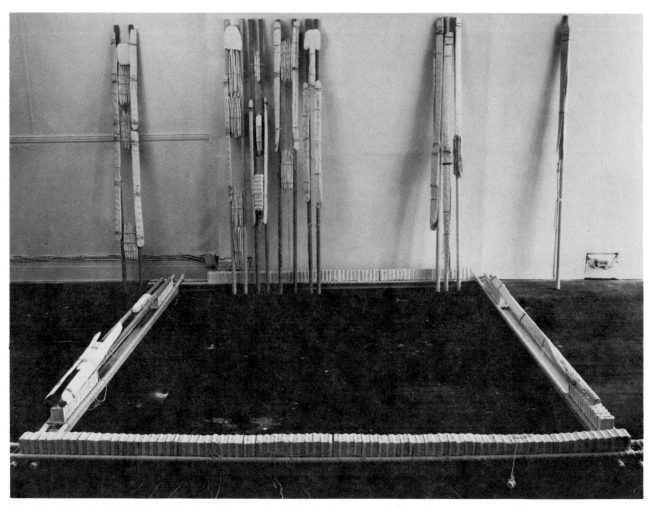

Altar *(in progress), 1976. Wood, handmade paper, masking tape, wax, and string.* 9—97 inches high; 2—109½ inches wide, 2—109½ inches long. Photo Yvan Boulerice.

JOSEPH ZIRKER

For many years, it has been my pleasure and privilege to witness the growth and development of works in many media by Joseph Zirker (b. 1924). He is one of those rare beings who takes nothing for granted; who tests what has been done before, but in his own way; who is so technically inventive, it is difficult to keep pace; who was a master lithographer early on; and whose association with papermaking and with papers and prints have had a long history.

In writing to me about the approach used in *Squashed Assemblages,* Zirker described the photo as ''a monotype printed on a paper sheet that sandwiches twine and a variety of cloth during the paper-making process. While the paper is still very damp, the top layer is selectively torn away, revealing the desired areas of cloth and twine. Finally, the whole is treated as a special sheet of textured paper and a monotype is printed on its surface—including the cloth and twine.''

Squashed Assemblage, #5–77, 1977. Handmade paper, twine, cloth, and inks. Approximately 15 x 18 inches. Photo Lon Otterby.

9. RECYCLING AND UNUSUAL PAPERS

Slowly the Bible of the race is writ,
And not on paper leaves nor leaves of stone;
Each age, each kindred, adds a verse to it,
Texts of despair or hope, of joy or moan.

—James Russell Lowell, 1849

As noted in previous chapters, making handmade paper is as easy as flying without fear—if you are a bird. Given some pulp, plenty of water, and a sieve or a screen (the equivalent of a mould and deckle), you can make an adequate sheet in your kitchen with simple tools and ordinary household appliances.

Now, suppose you are a painter of watercolors, a printmaker, a designer, or an illustrator who uses good quality matboards and imported, beautiful, mould-made or handmade papers for your work. Suppose, further, that you, as a sensitive user of fine papers, find yourself with a multitude of scraps of expensive papers and leftovers of museum-quality matboards. Or, perhaps, through circumstances beyond your control, you find yourself your own best collector of unsold, original prints pulled on a variety of handmade and mould-made papers. Will it break your heart to discard these remnants, burn them, put them in the garbage? Is there a way to salvage these odds and ends of paper and board? Can you recycle them?

RECYCLING SCRAPS OF PAPERS OR BOARDS
Although you can recycle any kind of paper and matboard, I would not recommend (your first time around) that you expend your time and energies on less than pH neutral, good quality materials.

Basically, there is one problem: how to convert, transmute, recycle paper or board back to pulp. The following is one method.

1. Cut or tear the paper or board to be recycled into 1 or 2 inch rectangles. At this stage, do not intermingle matboard and paper indiscriminately. I suggest you try, first, to recycle good quality paper scraps.

2. Soak the scraps in a plastic garbage pail (or a smaller plastic vessel if you do not have a large quantity of material to recycle) filled with pH neutral hot water. Drain the water, and repeat this step four times daily for at least two days. (Obviously, the water will be cool each time you are ready to drain the pail.)

3. Remove the scraps from the garbage pail, and put them through a meat tenderizer or an equivalent kitchen device. Try to avoid splashing water all over the working area at this time—you will be doing that regularly in the future.

4. Use a hydropulper. Or—if you do not have one—run the "tenderized" scraps in short bursts through a high-speed kitchen blender with about 95% water and 5% wet scrap. If your blender appears to be straining or having difficulty, increase the quantity of water and decrease the amount of scraps. Measure the proportions against the gradated scales usually found on the side of the blender, leaving some space for the mixture to splash upward. Do not be dismayed by the fact that you need so much water and so little solid substance. Are we not, as human beings, overwhelmingly composed of water?

5. Beat a new batch of pulp each time it resembles the photograph on page 41. As a very rough estimate, 1 pound of dry, recycled paper scraps will produce about 25 or so thick sheets of 16 × 22 inch paper.

6. Adjust the quantity of pulp to the water in your vat, and follow any of the suggested methods of sheet-forming that strikes your fancy.

BEVERLY PLUMMER'S FLOWER PAPERS
You will find Plummer's detailed procedures in her book, *Earth-Born Things,* proving she has followed the grand traditions of eighteenth century experimenters like Dr. Schäffer and others. Beverly Plummer speaks, works, and writes with honesty, candor, humility, and strength:

"I make sheets of paper from iris, cabbages, nasturtiums, rhubarb, gourd vines, cornstalks and clippings from the florist's floor.

"The plants are cut, chopped, boiled in lye, and then beat to a pulp in the blender. It's a savage process! Why do I do it? Why do I break down the plants and force them to knuckle under to me?

"I wonder if I'm dissipating some fiery destructive force from deep within my being? Or going into battle with nature knowing I won't get hurt too much? It could be I'm trying to create something beautiful to make up for the parts of me that I consider un-beautiful—finding a way to be reborn from ingredients I choose, rather than accepting those given me by my father and mother.

"Acts of creation are often savage. I believe the plants understand."

ELAINE AND DONNA KORETSKY'S BAST FIBER PAPERS
Here we have a combination of techniques and materials from the East and the West based upon research in papermaking techniques in Japan, Thailand, Taiwan, the Philippines, Indonesia, and Hong Kong.

"Briefly: we have been working mainly with a variety of bast fibers, much of which we grow ourselves —begonia, banana, sisal (agave plant). The banana plant has been especially versatile in papermaking, and I have six different varieties that we are growing and propagating. In addition, I am using various kinds of hemp from the Philippines and India, which I use alone or in combination with other bast fibers. My equipment consists of a small Hollander-type beater (1 pound size), two Lightnin' mixers (a proprietary name for a large commercial mixer), several book presses of different sizes, three fiberglass vats, and a whole group of moulds, all of which we have made, except for one English mould by the Amies Co.

"To process our plant fibers, we cook the cut-up plant material in a lye solution (1 teaspoon of lye to 1 gallon of water) in a large stainless steel pot. The plant material is cooked until soft and fairly mushy, then drained, and thoroughly rinsed, so all the lye is washed out. Then the material is beaten to a pulp—the length of beating time is entirely flexible, depending upon the amount of fibrousness or texture desired in the finished paper. We get a good range of color in the natural plant material—a whole range of off-whites, beiges, browns, and light greens. Or, the pulp may be bleached or dyed. After the sheets of paper are formed and water squeezed out in the book press, the sheets are brushed on to boards to dry. When dry, they are stacked up, and, if slightly wrinkled, are put back into the book press for a final light pressing.''

PETER T. SARJEANT'S JUNK MAIL APPROACH

In his book, *Handmade Papermaking Manual,* Dr. Sarjeant suggests that you can make 50 or so 8½ × 11 inch sheets of paper from one pound of recycled junk mail. Obviously, you should remove staples, cellophane, plastic windows (which will not make paper) and try to select papers, especially envelopes, with the least amount of printing on their surfaces. Also avoid wet-strength paper (towels, wash cloths, etc.) and use few, if any, brown paper bags. Unbleached kraft paper (brown bags) will lower the whiteness and brightness of your paper, will create odors in successive steps that may drive you out of the kitchen, *but* will strengthen your paper and satisfy your needs *if* used with discretion. Use it or not, as you prefer.

1. Tear or cut your junk mail into 2 × 2 inch pieces.

2. Place all the torn pieces in an enameled or stainless steel pot, cover with water, push down the paper, and add an inch or so more water.

3. Add a few tablespoons of household bleach such as Clorox, Purex, Topco, or the equivalent.

4. Place the pot on the stove, cover it, and bring it to a gentle boil.

5. Stir occasionally for about two hours. Do not be disturbed by the fact that the liquor (the resultant water from the boiling action) is now an ugly dark brown or red and quite smelly (though harmless).

6. Allow to cool. Break up any lumps with rubber-gloved or otherwise protected fingers. The smaller the scraps you started with, the simpler will this step be.

7. Remove pulp by the handful and squeeze hard into balls.

8. Place pulp balls in cheesecloth over a collander or sieve set in your kitchen sink.

9. Run tap water over the balls, and keep squeezing.

10. Set the squeezed pulp balls aside, and rinse the cooking pot.

11. Fill the pot with fresh water, add the squeezed pulp, and stir with an electric mixer. Be sure the pulp is diluted with sufficient water so the mixer does not jam.

12. Make pulp balls about the size of pigeon eggs (approximately a level tablespoonful of squeezed pulp). Each of them will make an 8½ × 11 inch sheet of paper.

J. N. POYSER'S EXPERIMENTS WITH KRAFT PAPER

In his book,. *Experiments in Making Paper by Hand* (see Bibliography), J. N. Poyser pays particular attention to making paper from wood pulps, wood straw, rope, bark, and some recycling. Mr. Poyser finds unbleached heavy kraft paper a most useful fiber material. In his own words, ''The repulping of heavy kraft paper without bleaching provides an easy source of good fiber material for pulp dyeing or for use in design work.

"For this experiment, it is not necessary to use unbleached kraft paper since any paper that is not plastic-treated, wax-impregnated, or too heavily printed can be used. I have used scrap writing paper or

white envelope paper in this experiment (greeting card leftovers). About 90% of all waste paper collected is repulped using a method similar to the one used here. A great deal of this waste paper goes into making paperboard.

1. Tear brown grocery bag paper in 1 × 1 inch or smaller pieces.

2. Weigh out 6.5 grams (.23 ounces) of the scrap brown paper, and place it in a quart-size Mason jar.

3. Add 100 milliliters (3.5 fluid ounces) of warm water and 100 milliliters (3.5 ounces) of 1% sodium hydroxide to the paper in the Mason jar.

4. Screw the cap on the jar firmly, and shake the mixture vigorously. The paper will be reduced to a lumpy pulp almost immediately.

5. Dilute the pulp to 600 milliliters (21 fluid ounces), and agitate for 5 minutes with a drill motor and paint stirrer, or 3 minutes in a blender.

6. Wash the pulp at least three times in a collander or sieve through cheesecloth. Place the collander in the kitchen sink, and turn on the tap. Wash until there is no trace of alkali, as indicated by the Alkacid test ribbon. The sodium hydroxide turns the ribbon deep blue, indicating a high pH value. (This is an important step since some dyes are decolorized in the presence of alkali or acid.) Sodium hydroxide in solution swells the fibers so they come apart more easily (the bonds are broken). The higher the concentration of sodium hydroxide, the more the paper swells, however, and the more difficult it is to remove concentrated sodium hydroxide.

J. N. POYSER'S FAST PULPING OF SPRUCE WOOD CHIPS

With the following preliminary words of caution, J. N. Poyser offered this experimental approach to fast pulping spruce.

"First, while I realize that sodium hypochlorite solution (Javex, etc.)

and sodium hydroxide (household lye) are generally available from local stores, they can be dangerous if not handled with caution. For example, when sodium hypochlorite is heated or comes in contact with acid or acid fumes, it can release highly toxic chlorine and chlorides (see Sax in Bibliography).

"In the lab, I keep solutions available as "anti chlors," such as 4% sodium bicarbonate or 4% sodium thiosulphate. Dilute acetic acid (vinegar) is easily available for sodium hydroxide solutions, although dilute hydrochloric acid is recommended by the Manufacturing Chemists Association, Laboratory Waste Disposal Manual (see Bibliography).

"The sodium hydroxide-sodium hypochlorite sequence is not an ideal way to delignify plant material, since there is some damage to the cellulose. However, the better methods are more dangerous for the novice.

"The experiments were really intended as school laboratory demonstrations, not for the production of pulp on a large scale, unless you can ensure good control of the environment, including protection for the people using the materials."

Thus, it is advised that you wear rubber gloves and protective eye-shields, make only small quantities of wood pulp, and exercise caution.

Preparing the chips. "Obtain a foot of 2 × 4 inch spruce board, then cut off 1 inch lengths. That is, you will have several pieces of spruce wood 2 × 4 × 1 inches in size.

"Split these 1 inch pieces along the grain at about ⅛ inch intervals, using a wood chisel or small hatchet and hammer. Discard the pieces with knots and blemishes.

"Break the pieces along the grain again so chips ⅛ × ⅛ × 1 inch are obtained. Straight-edged pruning shears are convenient to use for this step."

The process. "It is possible to make pulp in about a day and a half.

1. First, weigh out 20 grams (.71 ounces) of chips. Boil them for 5 minutes in 500 milliliters (17.5 fluid ounces) of water, drain them, and add 200 milliliters (7 fluid ounces) of 1.0% sodium hydroxide solution.

2. Simmer the chips at low heat (90° C. or 194° F.) for two hours. A 600 milliliter Griffin Pyrex beaker is very convenient for this operation, although a Pyrex or stainless steel saucepan can also be used.

3. Wash the chips with running water, and then transfer them to a quart Mason jar. The liquid will have turned a dark brown.

4. Wash away this liquid with water after covering the mouth of the open jar with a 6 inch square of polyester, using an elastic band. Fill the jar with water, and then place it on its side so the liquid drains away. Refill the jar with water, and repeat the process three times.

5. Add 200 milliliters (7 fluid ounces) of sodium hypochlorite solution, and allow the chips to soak for 8 hours at room temperature. Cover the open bottle with Saran Wrap held on with an elastic band.

6. Next, wash the chips three times; shake vigorously about 100 times in a bottle ⅓ full of water and chips.

7. Place the pulp in a second bottle, add 200 milliliters (7 fluid ounces) of 1.0% sodium hydroxide, and simmer for another 2 hours (90–100° C, about 212° F.) in the beaker or saucepan.

8. Wash the chips as before, and add 200 milliliters (7 fluid ounces) of sodium hypochlorite solution. Soak for 24 hours, wash, and shake as before.

"The pulp yield will be about 35% of the 20 grams of chips with which you started the experiment. You can expect to make four small sheets of paper, which have a total weight of about 7.0 grams A.D. If the pulping is started one morning, the pulp will be ready for use the next evening; hence it is convenient to prepare during the weekend.

"The fast pulping has been used on Spruce wood, Manitoba Maple, Weeping Willow, and Balsam Fir with success. In the case of Jack Pine, an additional two hour boiling in 200 milliliters of 1% sodium hydroxide followed by another 24 hour treatment with sodium hypochlorite solution was required, perhaps due to the high resin content of pine.

"Where small trees or branches were used, ½ to 1 inch discs were cut from the wood across the grain. The bark was peeled off using a jack knife, and then the wood was split using a wood chisel and pruning shears.

"When you make paper from this pulp, you will notice that there are tiny wood pieces in the sheets. These are called shives in the paper industry and are usually removed from the pulp before it is used to make paper. The shives will give your handmade paper a certain distinctive appearance."

ON JAPANESE HANDMADE PAPERMAKING

The cottage industry approach to handmade papermaking is still so strong in Japan that—it is believed—one village will keep from its neighbors the "secret ingredient" that makes its paper "different." What might be regarded as heresy and untenable, from a western paper scientist's point of view, may be admired for its esthetic considerations by members of other cultures.

Many treatises have been written about the history and practice of Japanese handmade paper, *tesuki washi.* These beautiful papers, by and large, are made by farmers and their families during their off-season, when snow covers the ground and the water is bitter cold. These papers are also made by superb craftsmen.

The 1976 United States exhibition of the handmade paper of Eishirō Abe, one of Japan's Living

(Right) Eishirō Abe examining dried kozo. (Below) Washing the kozo after boiling. (Opposite Page Top) Eishirō Abe forming a sheet of paper on the mould. Note the difference in design between the eastern and western mould. (Opposite Page Bottom) Abe examining papers. All photographs Kazuniko Ohori.

National Treasures honored for his papermaking techniques with *gampi* fibers, revealed the breathtaking beauty, strength, and variety of one master's work, plus the particular attitude of one nation toward its master craftsmen.

Most Japanese paper derives from the barks of *gampi, kozo,* and *mitsumata* trees. The bark of *gampi* (which grows wild) is stripped immediately after the tree is cut; *kozo* and *mitsumata* (cultivated widely throughout the country) are cut into bundles and then boiled or steamed to allow the bark to be removed by hand. The outer bark—*kurokawa*—in all these trees is black, and is pared away by knife-wielding women to reveal the white inner bark, *shirokawa.* The white strips of bark are either stored and dried or washed in clear-running, icy river water to remove any remaining dark particles.

Next, it is boiled for several hours in open vessels with about 10% of wood ash solution or other alkali, until the fibers are easily separated between the fingers. Several days of rinsing and washing in a clear running river are necessary before the fiber is literally beaten on a stone slab by two workers wielding heavy hardwood clubs.

Two things are unique in the making of Japanese handmade paper: (1) a viscous substance called *neri* (a vegetable starch obtained from the root of the *tororo-ai* (*Hibiscus manihot Medikus*) is added to the vat of pulp to slow the flow of stock through the mould and deckle and to disperse the long fibers; and (2) the Japanese mould, *keta,* is made of well-seasoned cypress, while the laid screen or *su* is made of thick bamboo laid lines tied with silk or horsehair chain lines to form a very flexible screen in one direction and a very rigid one in the other.

The vatman in Japan, seemingly, scoops his pulp on his mould as do Westerners; some eastern deckles are closed by individual pieces of wood held firmly with each thumb (note that Abe's mould and deckle

Cutting paper strips for weaving—this technique should challenge all paper-makers and printmakers everywhere. Courtesy Eishirō Abe. Photograph Kazuniko Ohori.

are of different design); but, there is no vatman's stroke to witness. Instead, there is a series of multiple shakes and strokes, more than one dipping, and a throwing off—at the very end—of the unwanted slurry. The result? As rich and varied as the personalities of the thousands of craftsmen who engage in making handmade paper in Japan.

One final observation: no felts are used in pressing the posts of paper. Each newly formed sheet is rolled off the removable laid screen and placed directly on top of the previous one. Heavy weights or a screw press are used to press the block of paper, which, as if by magic, separates into individual sheets after 24 hours under pressure! Final drying of the sheets may be on boards placed outdoors.

HANDMADE PAPERMAKING IN KOREA

One of the significant outcomes of the First Handmade Papermaker's Conference held in Appleton, Wis-

consin (home of Dard Hunter's magnificent collection of paperiana) in 1975 was the free interchange of information among a widely disparate group of loners who were brought together through the efforts of the indefatigable Joseph Wilfer.

Himself the proprietor of his own handmade paper mill (among his many accomplishments), Wilfer managed to create an atmosphere of good fellowship, pride of craft, the sharing of some secrets, camaraderie, and the worst breakfasts I have eaten in years.

There were—in addition to representatives from the several established handmade mills—engineers, technicians, consultants, and other experts in the white art of papermaking, plus some artists, many of whom gave presentations of their paper works in slide form or through performance.

Established and beginning papermakers from Canada, Latin America, Japan, and the United States mingled freely, exchanged

Asao Shimura's Short Story of Harukiri

Yamada Toshiharu (b. 1950), Japanese apprentice of papermaking, Tim Barrett (b. 1950), a Fulbright student of *washi,* Japanese handmade paper, and I flew to Seoul from Tokyo International Airport, two and a half hours. This is our first visit to Korea. A friend of Mr. Yamada's, a Japanese, welcomed us at the airport, and then took us to the downtown by taxi. So we had no trouble in Korea. Tim had been learning some Hangeul (the Korean language) before the visit, and talked to the cab driver: "I am an American, and they are Japanese." He can speak a little, but can hardly understand what people reply.

"Now we were in a lobby of a hotel in Seoul. It was hard to believe that we had so few language troubles. In Mexico (in 1975) I stood everything until I went to the capitol to visit my friend, not eating and sleeping, just sitting on a bus—48 hours' drive from San Diego. I had no Mexican currency and knew no Spanish words. But during that time I studied Spanish a little. Anyway, a little language goes a long way.

"The first evening in Seoul we met several Koreans, and Mr. Kim Ok Suck checked where paper is made now. One place is Harukiri near Sam cheog—they make paper in the old way. It sounds very interesting.

"We can go to Sam cheog by bus on the highway about five hours from Seoul. Sam cheog has a population of almost 50,000; it is a fishing town. We asked anyone who looked like he spoke English, "Where is Harukiri?" One person said, "Harukiri is five miles from here; you can go to Sam ho sa temple first, and then walk." He took us to an inn at last. The next day, we went to the temple by taxi, and walked on in the heavy rain. Got wet to the skin, and found a bridge that was under water so you could not pass. We gave up that day and returned to the inn.

"The next day we had a rest and went shopping at Gang reung. At a market we happened to see papers for sale. We asked, "How much?" and "Where do they make these papers?" They told us Yang yang, Miro, and Sam cheog, etc. The price is between twelve and twenty cents for a sheet of 2 × 3 feet. We now had a clue that they are making papers somewhere nearby. We decided to try again to go to Harukiri on a fine day.

"Can you imagine how they make paper by hand in the primitive technique? The tools are as follows:

Wooden vat, 6 × 6 feet

Wooden bat, 3 feet

Bar, 4 feet

Bamboo screen, 2 × 3 feet

Two logs, 3 feet each

1 deckle

"On Sunday we tried again to go to Harukiri. It was a lovely day, and many hikers were out. We went to the temple first and began to walk up. We came to the bridge and could cross. An hour later we came to a big waterfall. Chamise (a tea stall) is there. We asked people, "Where is Harukiri?" but they did not know. Next time we asked a hiker who could speak Japanese, and he told us that there is a papermaking place nearby called Shin heung ri. And Harukiri means a place where one can go and return in a day—an imaginary place. Shin heung ri is a small village with 163 houses. We got there by taxi, and found a mill and a house of papermaking.

"We visited the papermaker. His working studio was outdoors beside a stream. He was cooking kozo with caustic soda, washing it in the stream, and beating it with a bat. The method of papermaking is amazing—I never expected this style. It's almost Japanese (Japanese papermaking way, as against western style), but the position of the screen and deckle is unique—rectangular—different from the Japanese. They do not use any upper deckle. They make paper with a screen and only a lower deckle. The screen is put above the deckle, and the top of the deckle is put on a log, supported only by both hands. The papermaker first scoops toward himself and then right and left. This operation is repeated several times. The screen is carried to a pile of sheets and pressed with a log. Drying is done on the floor of Ontol room (room with floor heating system). Tororoaoi is used as mucilage.

"Koreans made paper by hand in the old days, and they still do except for cooking. Cooking with lye of ashes and hand beating makes a long-fibered, tough paper, I think. These papers made in the Korean style are used for shoji (paper windows), and rough surfaces. We visited some other places, Miro, Je cheon, Jeon ju, and Kyeong ju. In 1976, there were 226 houses of papermaking in Korea (by Mr. Kim Yeong Yeon in Weon ju city). In Japan, there were 784 in '76 and 741 in '77. The number of papermaking houses is actually decreasing both in Korea and Japan."

addresses and ideas, listened to technical discussion, attended seminars, enjoyed movies on handmade paper and, except for the sound of a violin played at curious times, harmony reigned throughout the proceedings and allowed and encouraged the birth of the *Friends of Paper*.

The above is a way of introducing Asao Shimura, a fellow papermaker at the Conference, who wrote me about Korean papermaking and his recent trip in the company of papermaking friends—to research present-day methods used in that country.

A NOTE ON EAST INDIAN HANDMADE PAPER
"This mill (Sri Aurobindo Ashram in Pondicherry, India) uses about 5 tons per month of rag—mostly *bagasse* (sugar cane)—gunny, and employs about 130 people in the papermaking factory and about 40 in the stationery department. The paper is air-dried and sold mostly by the ream in India. We make water color-drawing paper as well as an extensive variety of both pale and bright colors and a few novelty papers. The stationery department fabricates pads, envelopes, and a great variety of items."

DIEU DONNÉ'S APPROACH TO TRANSLUCENT EFFECTS
Bruce and Susan Wineberg, proprietors of Dieu Donné Press and Paper, offered this advice on "transparentizing" to those who have access to a beater:

"While overbeating will produce a transparentizing effect, the addition of proprietary chemicals will definitely yield better results. I recommend the proprietary sucrose acetate iso-butyrate, which gives the paper somewhat the disposition of waxed paper or glassine.

"Perhaps it's best to start with a metaphor: that any transparentizing medium is like a paint or ink, something to brush on, play with, create with. There really is no one recipe, no gallons per etc., no fine

measuring. Its use is superimposed on the paper, a secondary application. One ounce of the stuff will produce different effects on different papers, although in shorthand it may simplistically be said that one ounce in the beater will transparentize a small, 8 × 10, sheet. We use it to transparentize only parts of a given sheet, usually trying to bring out more clearly an area inside the sheet that either seems interesting or has in it something especially couched. We have couched in photographs between two white sheets, as a simple example, and painted with transparentizing medium the specific area of the photo, leaving the remainder of the sheet untouched.

RICHARD CLIFFORD'S FERN-CATTAIL PAPER
Still another approach to making paper from free raw materials is provided by Richard Clifford, who in addition to fern-cattail paper makes handmade sheets from Johnson grass, sunflowers, celery, cucumber, yellow onions, thistles, mimosa, lotus, lettuce, cabbage, alfalfa, plus rag and plant fiber combinations, bleached or unbleached, dyed or natural.

1. Gather cattail when green, stack, and let set for two days.

2. Cut and run the cattail through a hammermill, hydropulper, or a blender.

3. Add 1 ounce of alum per gallon of wet pulp in storage; add 1 cube of yeast per 5 gallons in storage.

4. Mix well, and use within two weeks.

5. Place a sufficient amount of pulp in the vat—the amount depends on desired thickness or preference in dipping.

6. To the vat water, add 1 ounce of polyvinyl alcohol and 1 ounce of gum arabic per 10 gallons of liquid.

7. Obtain dried ferns from your local handicraft supply store, or pick them fresh.

8. Soak the desired ferns.

9. With a helper, stir the pulp in the vat, dip the mould and deckle down in and pull up to water level, *but not out of the vat* to allow a floating surface. (If you are a beginner, stretch an organdy or polyester curtain sheer over desired frame size and staple edges for a mould; make identical frame size for the deckle.)

10. Have your helper place the wet ferns or flowers as desired and pat them down into the pulp, allowing the pulp to spill slightly over the edges, but holding the ferns or flowers securely to the floating surface.

11. Pull the mould and deckle straight up from the vat, letting the water drain. Tilt the mould and deckle to allow more water to drain away.

12. Remove the deckle; place the mould in the sun to dry (or place in a protected area inside). Let the sheet dry on the screen of the mould.

13. When dry, peel the sheet from the mould.

CHARLES R. STRONG'S MOULDLESS PULP PAINTINGS
1. Place a piece of strong, heavy, galvanized wire screen in a shallow tray; add a soft, fine plastic screen (similar to window screen) on top of the galvanized wire.

2. Gently, pour in just enough water to cover the screens.

3. Using a variety of means for deckles (aluminum or other "dams," clay, or embroidery hoops), add various colored pulp forms to the screen as you pursue your visual goal.

4. Be certain to control the water level in the tray so the several shapes "settle in" but do not float away.

5. When you are satisfied with your image, lift the screens carefully and move them to a plywood

board. Now, place a damped felt on top of your image and another plywood board on top of the felt.

6. Hold the boards together tightly, flip them over, and remove the plywood board and the screens—carefully.

7. Either couch a larger sheet on top of all of your colored forms, or place a dampened felt down on the top side of the pulp. Replace the plywood board, and press out the water to complete the process. No paint is added at any time; the painting process and the papermaking process are one and the same.

In 1976, Strong was recycling clothing, high quality 100% cotton, Irish linen dresses, pajamas, and policeman's shirts. The following year he obtained his pulp already dyed from commercial sources.

THE VACUUM BREAKTHROUGH
In September, 1977, I called Harold Persico Paris to ascertain the present state of his radical, innovative approach to papermaking. Quite by chance, it appeared that I had called him just hours after he had solved certain technical difficulties that had prevented him from realizing those visionary qualities he seeks in his three-dimensional works. The solution involved vacuum approaches.

Several weeks prior to this, print-maker and papermaker Joseph Zirker had visited Charles Hilger's studio, drawn a version of the illustration and tried to explain the vaccum apparatus Paris showed Hilger:

"It (the apparatus) consists of a vacuum pump attached to a canister by a tube (plastic and transparent) that, in turn, has many plastic tubes leading to the underside of the working table. Each clear, plastic tube is attached at both ends to airtight fittings. As I observed the operation, a sheet of pulp is couched on a screen material (window screen or fine mesh plastic) laid out on the table surface. Next a piece of PVA [plastic], which is the thickness of the stuff dry cleaners use, is placed over the couched pulp and is sealed merely by wetting the table surface with water along the edges covered by the PVA.

"When the pump is started, the thin, transparent PVA material very quickly stretches tight (within a minute or two) along all edges and contours of the pulp and/or anything else placed on the table. Water can be observed flowing through the many (perhaps a dozen) tubes leading from fittings under the table to the cannister. A gauge fitted to the pump indicates inches of mercury, and the pressure soon reaches between 25 and 27 inches, or approximately 14 pounds per square inch."

Returning to my phone conversation with Harold Paris, his voice —filled with the enthusiasm of a youngster enjoying his first lollipop —was infectious as he described how his improved system allows him to create large-scale paper works in three-dimensions: "Sheets up to 6 inches thick that hardly can be drilled through," other works 5 × 10 × 1 foot in relief, and so on. His ideas and projects spilled out faster than my ability to record them.

It appears that Paris found that a single, large hose leading to the vacuum pump through the cannister is infinitely more effective for high-relief works than a number of small, clear plastic hoses. The danger of leaks is reduced, and this accounts for his vacuum readings of 30, which provides almost instant dryness of his vacuum-formed pieces.

This merely suggests another mode of forming flat sheets and low-relief collages. The inventive reader who demands high-relief devices, will find that an old refrigerator will yield a vacuum pump at a minimum price. Whatever, for those concerned with this approach, this very brief description allows you a running start on a machine that may astound you. (See also pages 134–135 describing Peter Gentenaar's use of vacuum approaches.)

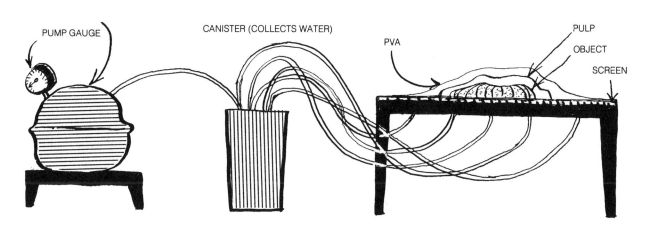

PUMP GAUGE   CANISTER (COLLECTS WATER)   PVA   PULP   OBJECT   SCREEN

The vacuum approach.

10. CAST PAPER

Then he that made, for us a Paper mill
is worthy well, of love and worldes good will,
And though his name be Spillman by degree,
yet Help-man nowe, he shall be calde by mee,
Six hundred men, are set a worke by him,
that else might starve, or seeke abroad their bread,
Who nowe lives well, and does full brave and trim,
and who may boast they are with paper fed.

—Excerpt from a poem by Thomas Churchyard, 1528

Sculptors, painters, and printmakers have rediscovered cast paper as a most expressive three-dimensional, permanent medium allowing the utmost freedom, monumentality or microstructures, and whiter-than-white or polychromed works to accommodate any and every mode of plastic vision.

The only requirement for the artist, assuming the creation and manufacture of a flexible or rigid mould or moulds, is great quantities of pulp, which may be purchased fully beaten—as half-stock or linters—or made from rags, linters, plant materials, etc. by oneself.

Though we are somewhat afield of our subject, it seems appropriate to describe briefly, negative and positive mouldmaking. Thus I requested one artist-papermaker, Suzanne Anker, to give us an insight into her working methods.

NEGATIVE MOULDS
BY SUZANNE ANKER

Building a Construction. "A construction is built using materials, such as cardboard, styrofoam, fabric, string, plaster, etc. The construction can simulate a collagraph plate or can be very high in relief (mine are generally about 3 inches deep). Essentially what is made is a relief collage or assemblage, which is then coated with several coats of varnish or verathane. The coating seals the construction and aids in releasing the latex rubber mould. The sealed construction is then sprayed with silicone spray, to allow an even cleaner release."

Mould-Making. "Though a plaster mould can then be taken directly from the construction, to safeguard myself against broken plaster—and to further manipulate indirectly the image I use an intermediary procedure of a latex mould.

"I brush rubber in liquid form on to the construction and allow it to dry. Heat lamps may be used to speed the process. I brush another coat on, and again allow it to dry. Then I place a layer of cheesecloth

on top of the dried latex coats, and brush another coat of latex on top of the cheesecloth. Altogether, five or six coats of latex are used. In about 24 hours, the latex will cure, and it will peel away from the construction.

Plaster casting (secondary mould). "If the original construction was not high in relief, there is no necessity to make a plaster mould. The purpose of the plaster mould is to reinforce the latex, which is limp and skinlike. For constructions high in relief, plaster moulds are necessary in order for the latex to attain its desired structure.

"I place the original construction and latex mould face-up on a table. A wooden frame, or clay, is used to form a boxlike structure into which the plaster is poured. I mix the plaster into cold water until it forms a creamy mixture. Be sure not have air bubbles present when you do this. The creamy mixture is sprinkled on to the mould and then poured, in a light even coat. Successive coats are used, as well as burlap or wire enforcement until the plaster is 1½ to 2 inches thick. I cast the plaster casts thick because, in the process of hand felting, pressure will be put on the mould. The thicker plaster will not break under the hand felting process.

Casting. "After all the moulds that are being used for the process are thoroughly dry, the paper casting process begins. The plaster mould is laid on a table with the latex mould on top of it. Since it is the latex mould that will hold the paper pulp, it is necessary to spray the latex with Krylon Crystal Clear. Through my experience, Krylon brand is the only one of its kind that does not impart a plastic skin upon the paper. The acrylic spray helps keep the latex mould water-insoluble.

"I then place or pour paper pulp over the surface of the latex. After a uniform or desired thickness has been reached, as much excess water as possible must be re-

moved with large sponges. Starting on the perimeter of the piece, I gently extract the water. Going in both horizontal and vertical directions, gentle pressure with the sponge should be used. This process is repeated, increasing the pressure after the first felting, until as much water as possible has been extracted. The piece is ready to be dried either naturally or force-dried with fans. About three-quarters through the drying process, I place weights (1 × 2 inch duct-taped sticks) around the perimeter to avoid warping. When the piece is dry, I lift it out of the plaster mould with the latex skin still attached. Turning the piece over, I remove the latex gently. The resulting paper piece will resemble the original construction; the latex mould is negative.

Color. "Color may be added in a variety of ways. Acrylic paints, fabric dyes, or watercolors can be added directly to the paper pulp to tint it directly. Moulds may be cast using a variety of colored pulps simultaneously. The colors will bleed into one another, causing a quite interesting watercolor effect—though if acrylic paints are used as a coloring agent, they will not bleed as much. However, you cannot exactly control the color in the casting process. Color may also be applied to the surface with watercolor and a brush, an airbrush, pastel crayons, or colored pencils. Applied color allows a high degree of control. I use color in my own work to highlight form as well as impart emotional content."

AGAR MOULDS

Agar is a colorless, formless vegetable gelatine obtained from marine algae; it is soluble in hot water, and appears to melt and liquify in it. It is a flexible mould which may be necessary in casting from forms that have undercuts; it usually requires a "mother" mould (plaster of Paris) to support it, as it is less strong than a gelatine mould. To make an agar mould, follow the in-

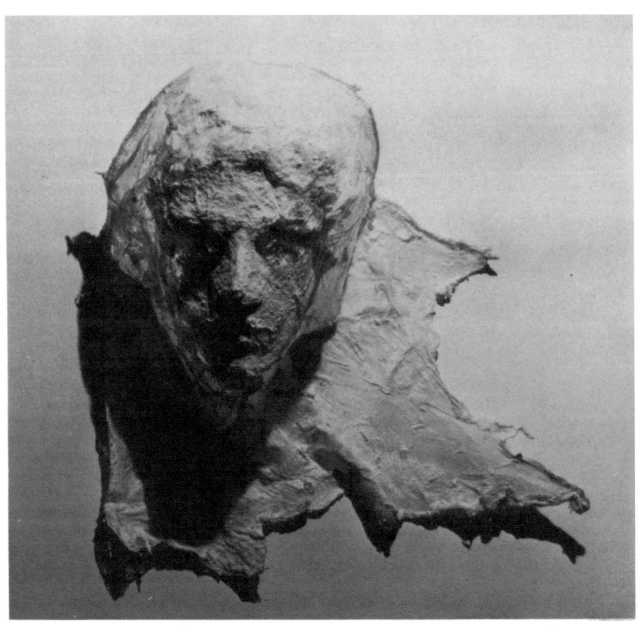

Zabie, 1976, by Garner H. Tullis. Pulp and gauze, 24 x 24 x 10 inches.

structions below—it may be used for casting objects in either wax or plaster (Butler, p. 27).

1. "To 6 ounces agar, add 1 quart water.

2. When thoroughly soaked, boil in double boiler, and add 3 pounds glycerine. Continue cooking until the water is well evaporated.

3. Add a few drops carbolic acid. Allow to stand until evaporation is complete.

4. The resulting rubbery mass should be cut into small cubes and is ready for use.

5. To use the cubes, heat, as before, until liquified, and—after cooling somewhat—pour over the object to be cast.

6. If you are casting in wax, harden the inner mould surface by coating with alum water before pouring hot wax into it.

GARNER TULLIS AND HIS PROCEDURE NO. 13

One of the innovators in the papermaking revolution, a self-taught bundle of extraordinary energies, a man who lives for each new experiment and stands for ideas and concepts that place him at the center of controversy and who, I believe, would not want to have it any other way, is Garner Tullis (b. 1939), founder of the Institute for Experimental Printmaking. If there is a way for the artist to benefit by what industry has discovered through large expenditures, Tullis has devised an equivalent for use by artists who come to San Francisco to work with him.

Ambiguous, universal, strong, frontal—these are terms that first come to mind in confronting a Tullis work in paper for the first time. Careful study of his work suggests new roads to search out the dignity of man, to discover a certain oneness within our individuality, to find within the vital image a burning desire to break out of the seeming serenity of the facade or mask worn by every man and every woman.

One of the many approaches used by Garner Tullis for creating his secret, enigmatic, mysterious, arcane, powerful cast paper pieces may have been created in the following manner. Faults and errors in my description may be attributed to the fact that I was engaged in work that only allowed me an occasional glimpse of Tullis, now and then.

1. First and most important, he pushed, pulled, modeled, cut, bruised, smoothed, attacked, caressed, brutalized, and, magically, "humanized," a rather large lump of clay to form a seemingly "universal" head.

2. The next time I noticed him was when he was draping a piece of cheesecloth soaked in Elmer's glue over the head and pressing it down in all the tiny hard-to-get-at places of the mask. The head was sitting face-up on a table top.

3. If memory serves, the head was still face-up on a table and Tullis was flicking, placing, pouring, arranging, rearranging cotton pulp on the gauzelike substance to transmute, the head from a thing of pedestrian materials to something of esthetic value.

4. I recall seeing infra-red heat lamps focused on his paper piece before I left the shop one evening.

5. The next time I saw the work, was in a well-known gallery in New York City, where it attracted, compelled, and hypnotized many hundreds of eyes one opening night.

11.
PAPERMAKING FOR SCHOOLS

The fate of the country . . . does not depend
on what kind of paper you drop into the ballot-box
once a year, but on what kind of man you drop
from your chamber into the street every morning.

—Henry David Thoreau, 1854

You have only to look at the eyes of the audience, be it composed of children from the age of 6 to 86, in the middle of a workshop on papermaking to truly comprehend the fascination, the deep-felt relationship that exists between them and the medium (handmade paper). Unspoken messages pass between, among, and from the individual demonstrating this magic phenomenon and those observing this contemporary shaman.

People of whatever age can barely restrain themselves (many do not) from instant participation with something that has been eluding them all of their lives. It appears as though papermaking awakens primordial memories, long-since buried, long-since carted off to the dustbins of human history and asks —demands would be more accurate—that we involve ourselves in this simple-complex process at whatever level our needs, desires, abilities to work with tools, materials, and equipment dictate.

Thus, this chapter is devoted to all of us—to the child that was and always will be in all of us.

Two things bear repeating: (1) no one will become a professional handmade papermaker by reading this or any other book, nor will anyone be immediately competent enough to produce sheet after sheet, ream after ream of the same quality, size, texture, thickness, and color—it takes years of apprenticeship to a master—and (2) cleanliness of all tools used, freedom from rust (the enemy of papermakers), cleanliness (pH neutral) of the water employed, cannot be overemphasized.

DARD HUNTER ON PAPERMAKING FOR CHILDREN

When first I learned that Dard Hunter had written a book titled, *Paper-Making in the Classroom,* my impatience with interlibrary loans could not be contained. No copy existed in Arizona. I wanted, desperately, to read what the *maestro* had to offer children with regard to rag papermaking by hand.

It is a slim book of 80 pages with a two-page index; more than half the book is devoted to a history of paper and modern industrial production of the same. A précis of the advice for making handmade paper offered to children, based upon a close reading of the book follows:

1. Find well-worn, white cotton rags (dresses, handkerchiefs, napkins, etc.) rather than linen. If new cotton

Robert Hauser of Busyhaus forming a sheet during a workshop seminar on papermaking.

cloth is used, first boil in water to which you should add "a little caustic soda." This helps remove the sizing or glutinous matter contained in the cloth.

2. Use a miniature beater (Hunter's term for a 5 to 20 pound Hollander-type beater that will accommodate that much dry material!). He states that several schools have already installed such beaters.

3. Fill the beater tub to ½ capacity with clear water. Raise the beater roll clear of the bedplate, and set the machine in motion.

4. Cut the rags in 3 inch squares, and feed them slowly in front of the beater roll. Thus, the rags will be drawn between the knives of the beater roll and the knives of the bedplate, and thrown over the backfall with sufficient force to allow them to recirculate around the midfeather and then back to the beater roll.

5. Do not adjust the beater roll so it is too close to the bedplate. You want to fray and draw out the cotton fibers, rather than cut them to bits. Hunter repeats the statement that long fibers make the strongest paper and short ones provide papers that are easily torn, though he does confess that the latter provide "clear and brilliant" watermarks.

6. How long do you keep the rags in the beater? Hunter's answer is equivocal, as it must be, and he suggests conducting tests.

7. If no beater is provided by the school, Hunter offers the following substitute.

Use a pair of scissors or a knife to unravel, fray, or draw out the fibers in the rags, or find an old-fashioned meat grinder. (Remember, this book was written in 1931; where in the late 1970s will you still find what was old-fashioned in the 1930s?).

Soak the squares in water, and feed them into the grinder very slowly. The only clue offered about the grinder is that its blade should not be adjusted too tightly.

Put the rags through the grinder two or three times.

Test in the usual way for proper beating. Watch for lumps and any foreign matter.

Now, place the stock in a large bowl or jar with sufficient water to bring it to the consistency of heavy cream. The pulp can keep for several days without developing mould or creating other problems.

Hunter describes the manufacture of professional laid and wove moulds and deckles of mahogany and also shows photos of small (postcard size) to large (13 × 10 inch) crude, student-made moulds and deckles of pine. Aluminum moulds are illustrated, and he suggests they are made for laboratory testing in the paper mills and available inexpensively. (They were.)

The vat suggested is the common wooden tub or galvanized metal tub used for clothes washing. It should be placed on a bench, so the top edge comes to the waist of the vatman. Hunter suggests a screw press, such as a bookbinder's press or an old table press. And, lastly, he states that you can acquire worn felts from a paper mill.

Watermarking, forming, couching, sizing, coloring, and drying are described in straightforward, traditional fashion. Paper, for example, is air-dried, in spurs of several

Young papermaker "at the vat."

sheets, on heavy cords or wooden poles in the classroom.

The only amazing statement in the book is that which reads, "Provided it is not desired to go into the making of paper extensively, the entire equipment for a school paper mill should not cost above two or three dollars." Heartbreaking?

THE ONTARIO SCIENCE CENTRE
On a single sheet of 8½ × 11 inch colored stock, the back of which contains an excellent bibliography, the staff of the Ontario Science Centre, in Toronto, Ontario, Canada has put together an attractive, approach to making handmade paper:

1. "Make yourself a wooden screen from scrap pieces of wood. (This is illustrated with a line drawing showing an 8 × 8 × 1 inch thick wooden frame.)

2. Staple fly screening tightly and smoothly across the frame. Use nylon fly screening—it's easier on the fingers and doesn't rust.

3. In a blender, mash up an old Christmas, birthday, or report card along with some potato peels, or carrot peels, or any vegetable fiber you happen to have around. You'll have to experiment.

4. Dump the mixture (called pulp) into a pail, tub, or kitchen sink filled about 4 inches deep with water.

5. Then grasp your screen in both hands, place it in the tub, shake it gently from side to side, and in a single straight motion lift the screen out of the tub. The water will rush through the screening, but the fiber will have evenly coated the screen. Remember, try to keep the screen level—if it tips, half your paper will be really thick, and half will be too thin.

6. Now, to get your paper off the screen. (This is the tricky part.) You will need some old newspapers laid on a table and an iron. Take your screen, turn it over, and place

it on the newspaper. Mop up the excess water with a sponge; then very, very carefully lift up the screen. Your paper will remain on the newspaper. Next step is to iron the paper dry by putting a sheet of newspaper over the new paper so the new paper doesn't tear. And that's all folks—good luck and good paper. Your friends will be amazed."

The information sheet is headed as follows: "Papermakers of the world unite. This is a simple recipe about papermaking, and how to go about it. We have been experimenting with paper for a while now, and it's been fun. If you have any ideas, or fantastic recipes you have come up with, please drop us

a line: Papermaking, c/o The Ontario Science Centre, 770 Don Mills Road, Toronto, Ontario, Canada."

It should be pointed out that the Science Centre is the proud possessor of a laboratory-size Hollander-type beater and that they have been experimenting with all sorts of local flora, linen, and cotton for a number of years. Finally, the papermaking area is always jammed with eager persons of all ages at each of the demonstrations each day.

A VARIATION OF THE API, INC. METHOD
This variation on an introduction to papermaking derives, for the most part, from a flyer published by the American Paper Institute, Inc.

Couching the newly made sheet.

Equipment Required. Collect the following:

1. A mould and deckle.

2. A basin (preferably enameled) that will hold at least 10 quarts of water.

3. 30 sheets of facial tissue (not wet strength).

4. Two sheets of blotting paper.

5. Liquid laundry starch. (One tablespoonful of instant starch in 2 cups of water will provide paper-maker's size.

6. An egg beater or blender and a rolling pin.

7. A household electric iron.

I suggest you first make a mould of four pieces of 1 × 2 inch thick scrap wood or wooden canvas stretchers. Select a rectangle of a size that will fit easily within the basin to be used. If you do not have a basin, use your kitchen sink.

For a sheet of paper 8 × 11 inches in size, you will need to construct a 10 × 13 inch frame or mould. For a deckle—to fit snugly over the mould and fence in the pulp—use a simple picture frame of appropriate dimensions.

Staple regular house screening of plastic, brass, or copper wire, or use polyester fabric over the mould. Stretch it tightly, as you would canvas over wooden stretchers.

Steps in the API, Inc. process. To make the paper, follow these instructions:

1. To approximately 10 quarts of water in the basin, add the starch mixture. Stir.

2. Tear sheets of facial tissue in 1 to 2 inch squares, and pop them in the basin.

3. Using an eggbeater or a food blender, reduce the tissue until it is fully pulped.

4. Grasp the paper-forming device or the mould and deckle suggested firmly. Hold it vertically to the basin of pulp, and dip the bottom end in the pulpy mass.

5. Bring the bottom of the forming device or mould and deckle toward your body. Quickly, shake the pulp back and forth and sideways to lock the fibers and form an even sheet. The water may drain so quickly that you will not have enough time to shake the pulp. However, if the pulp is distributed fairly evenly, proceed; if not, turn the face of the mould toward the basin and kiss off the pulp to try again.

6. Remove the deckle or screen filled with pulp. Place it between blotters, and remove the excess water by using the rolling pin over the blotters. Or, you may wish to place the mould in the sun, and allow the paper to dry on the screen.

7. Iron-dry the sheet with a warm iron (if you have chosen not to air-dry the piece).

8. The result should be your first sheet of recycled, handmade paper.

You may wish to put some lint from your dryer into the basin before forming your next sheet; it will provide a unique paper having multi-colored fibers running in, on, and through it.

As you acquire experience, you may wish to substitute filter paper, photographic blotter paper (they are both pH neutral), tiny pieces of old cotton rags, or scraps of good printmaking or watercolor papers.

Always add your well-soaked (overnight or longer) paper scraps or rags *to* the water in the blender (when using a food blender). It is better to use too much water than too little. Besides, if you attempt to macerate too much material at one time, you will burn out the blender motor.

ANOTHER KITCHEN APPROACH
This formula was contained in an interview with an artist-papermaker by Nancy Davis and published in 1976. There was a warning placed before the recipe: "Don't toss rags or other cottons into your blender; you'll probably ruin it. In following these directions, be alert for signs and smells of mechanical distress from your blender.

1. Soak 20 sheets of toilet paper or 5 sheets of paper towels overnight in 3 cupfuls of water.

2. Pour into a blender, and mix for five minutes, switching the machine on and off rapidly until you have a bulky, cohesive batter. It might be wise to soak a few extra sheets to use in case your batter turns out too thin at first try.

3. Time for creativity: add food color, dye, flower petals, wisps of yarn, or anything else your imagination suggests.

4. Pour the batter on to a sheet of muslin, linen, kitchen toweling, or felt that has been tacked to an empty picture frame about 12 × 13 inches. Let it drain—flat—for 3 to 10 hours. If you prefer thin to thick sheets, divide the pulp in half and proceed as above."

DR. J. B. LIEBERMAN'S APPROACH
In a major work on papermaking, Dr. J. Ben Lieberman (see the Bibliography) includes a section on "A Child's Way to Make Paper," which reinforces the compulsion most people feel when they first encounter handmade paper being made:

1. Cut old rags into 2 or 3 inch squares, and place them in a "slightly rough-surfaced bowl" with water.

2. Beat the rags (using any method in this book) to flay, fray, and separate the fibers.

3. Test the pulp from time to time for lumps or knots; when the test jar is milky white, your pulp is ready. (Waste paper or natural fibers may be used, as well.)

4. Place the pulp in a vat or tub with sufficient water to bring it to the consistency of heavy cream. Add starch if you want the paper to take

ink well; dye, if you want it colored; or nothing—as you choose.

5. Stir the slurry vigorously. Using a silk-screen and a frame (or any mould and deckle described in this book), dip it vertically into the vat of slurry; turn the mould and deckle to a horizontal position to catch the fibers on the surface of the mould; lift it out of the vat, still holding it level; give the mould and deckle a slight jerk to better lock the fibers.

6. Keep the mould level to allow all the water to drain. Now, either place it in the sun to air-dry (which will allow easy removal of the sheet, or couch it on to a dampened felt or wool blanket (see page 47 for proper couching procedures).

7. If you use the couching method with wet sheets of paper, place a felt over the first sheet; make another sheet and couch it on to that felt; cover this one with a felt, and repeat this procedure until you have a pile or post of wet paper sandwiched between felts.

8. Place the pile under a press of some sort to squeeze out as much water as possible. Separate the felts, and place the damp paper in the press once more (in a different order) to expel more water. Separate the sheets.

9. Hang the sheets over a pole or a line to dry—or lay them out flat. If the former method is employed, use three or four sheets together to avoid too much curl.

DR. LIEBERMAN'S VACUUM SUCTION PAPERMAKING BOX

Although this particular tool was designed for allowing handmade papermaking to become a cottage industry by taking the skill out of forming sheets of paper with a mould and deckle, it may prove to be of use in a classroom situation, or in any situation in which you would like to see perfectly formed sheets manufactured one after another, without end.

Building the box.

1. Build a box of plywood, no less than 12 inches high, the inside dimensions of which will be the size of the desired sheet.

2. About 3 inches from the top, fasten a 1 inch wooden ledge on the inside of all four walls on which your mould will rest.

3. Build a mould to fit snugly inside the box (use any of the methods shown between pages 68 and 70), and affix rope, nylon fishing line, or equivalent handles to the wooden frame of the mould to allow for easy removal from the box.

4. Be certain that the mould frame is secure (almost watertight) against the insides of the box. Use rubber gasket material on the inside walls to make certain this is so.

5. Design your box according to A or B in the illustration at right. In either design, the slurry (which will be in the box) must be allowed to be evacuated instantly.

6. Using one of the proprietary waterproofing substances on the market today, waterproof the inside and outside of the box and the mould.

Operating the box

1. Fill the box with slurry (some persons may prefer to use water at this point) until it covers the screen in the mould that sits on the ledge provided for it, inside the box. Place the handles of the mould so they dangle outside the box. Be certain the bottom is closed.

2. "Pour a measured amount of slurry into the top of the box, stirring it as necessary to make sure the mixture is evenly distributed." (Dr. Lieberman, *op. cit.,* p. 82.)

3. Without jerking the box, release the slurry *instantly*. Doing this properly will create a vacuum when the bottom slurry rushes out sucking the water down through the mould screen sufficiently hard to felt the fibers on its surface and create a perfectly formed sheet.

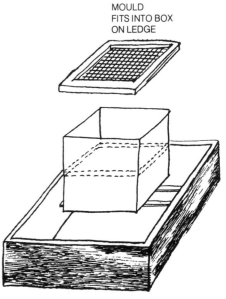

MOULD FITS INTO BOX ON LEDGE

Dr. Lieberman's vacuum suction papermaking box.

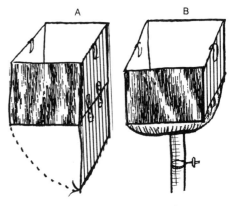

A   B

Details of the papermaking box, bottoms A and B.

4. Remove the mould and couch the sheet.

Problems to solve. The problems can be many and varied:

1. The mould must truly fit tightly against the sides, yet should allow for easy lifting from the box. The former relates to the vacuum to be created; the latter for your convenience.

2. Find a strong, appropriate handle for your mould that will not constantly be in your way.

3. If too great a vacuum is created (you will be aware of this phenomenon without further comment), your alternatives are either to move the mould rests down lower, or build a smaller box.

4. With regard to using water in the box instead of slurry, it may be regarded by some individuals as wasteful; if you use slurry, it can be reused.

5. The B solution in the illustration on page 181 may prove to be the better solution for larger sheets of paper.

Though the box was not field-tested when Dr. Lieberman wrote his book he was certain it would work and requested data from those who used it, with what fibers, under what beating conditions, and with what success.

In 1966 J.N. Poyser of Pointe Claire, Quebec published the first edition of his *Experiments in Making Paper by Hand,* which utilizes a vacuum suction papermaking box. Mr. Poyser suggested that *his* solution was but a variation of the Canadian Standard Freeness Tester and the British Sheet Machine.

ALTERNATE MATERIALS FOR PAPERMAKING

Besides wood, which is used as the raw material for the vast majority of machine-made papers produced in the world today, we know that we can use almost any plant that grows anywhere, provided it has some cellulose. And, if we mix any plant material with some cotton linter stock, the world of the unknown is ours.

In some of my classes, either in Canada or in the United States, students have made handmade paper from carrots, grass, straw, banana peels, old fish nets, artichoke hearts, hemp rope, horse chestnuts, various weeds (still unknown), seeds, lint from clothes dryers, silk, sawdust, moss, leaves of all the trees within a 15 minute walk of certain campuses, pine cones, potatoes, roof shingles, marsh mallow, nettles, oak, willow bark, yucca, and sundry desert flora.

The materials were either obtained locally or sent to students by interested friends. As has been reported by all experimenters with alternate materials for papermaking, there were no failures when some cotton or linen was employed in the batch, no matter how small the quantity. In many instances, though, it was possible to make paper entirely of the alternate material.

Professor Feng-jyi Chang of the National Chunghsiung University in Taiwan has made special purpose high quality paper from the pineapple leaf, taking advantage of the fact that about 15,000 hectares of pineapple are cultivated on plantations in his country.

Now, *you* try something else.

HUMAN RESOURCES FOR PAPERMAKING KNOWLEDGE

In addition to the names of practicing artists mentioned in this book, there are numbers of persons, groups, and organizations currently offering lectures, lecture-demonstrations, workshops, and seminars on hand papermaking in North America and on every other continent in the world. All of these persons are knowledgeable, enthusiastic, resourceful, and guaranteed to entice students of any age into the wonders of papermaking.

To mention them all by name (an impossible task) would surely encourage omission of one or more excellent speakers. Consequently, I suggest, if you are interested, you inquire of local galleries, colleges, museums, universities, and critics on your newspapers for up-to-date listings.

There is one individual whose presentation is unique, however—Arnold Grummer, former Curator of the Dard Hunter Paper Museum, actually transports a large, transparent, papermaking machine with him as he makes his rounds, which was crafted at the Institute for Paper Chemistry in Appleton, Wisconsin. The machine, in addition to his unusual visual material and lecture on hand papermaking, appears to fascinate audiences of all ages.

Many, if not all of these good people, carry portable equipment and supplies—including slides, films, moulds and deckles, pulp, hydropulpers, fully beaten pulp, linters, and small hydraulic presses —to supplement their tasks. All offer qualitative visual and aural information and workshops on papermaking by hand.

West Hall Underground, the inaugural class in papermaking at Arizona State University, 1977. Photo Tom Morrissey.

12. ABOUT PAPER

When you sell a man a book you don't sell him just twelve ounces of paper and ink and glue— you sell him a whole new life.

—Christopher Morley, 1917

In the long view of history, it was only a few moments ago that the paper wasp's unique approach to papermaking was noticed by man in general and by a gifted French scientist, René Antoine Ferchault de Réamur (1683–1757) in particular. But our story truly began in China more than two thousand years ago.

When archeologists unearthed scraps of very thin, yellowish, vegetable-fibered paper—in a tomb at Pa-ch'iao in Sian, in Shensi province—believed to have derived from the early period of the western Han dynasty (202 B.C.—9 A.D.) they, like it or not, destroyed a long-held and honored view put forth earlier by Sir Aurel Stein and Dr. Sven Hedin. Those two men, in their finds in Mongolia and Turkestan strongly suggested that paper was not in common usage until the second, third, and fourth centuries A.D.

Fifteen years before the Shensi province find, Professors Lao Kan and Shih Chang-ju of the Academica Sinica discovered what they believed was "the oldest paper in the world in the ruins of a watchtower in Tsakhortei, south of the Bayan Bogdo Mountains, in the modern Ning-hsia area." (*Papermaking, Art & Craft*, LC, p. 9.) One can read about two dozen Chinese characters on this puckery ball of coarse paper, which may have been buried by chance about 109 A.D. during an attack by the western Hsi-ch'iang tribe.

Other experts suggest that the origins of paper extend back two to three hundred years earlier. Whatever, we will leave them to their scholarly disputes. Suffice to say that teams of Chinese archeologists, since they are once more in the field, may move mountains of earth to shake the foundations of many legends about the origins of many things.

FROM TS'AI LUN IN CHINA

It has long been suggested that paper, as we know it, was invented in 105 A.D. by Ts'ai Lun of Lei-Yang,

north of Canton in South China. Ts'ai Lun was Privy Councilor in the court of the Emperor Ho Ti.

The story of the invention of paper has been embroidered freely throughout two thousand years to confound, confuse, contradict, and amuse historians—for instance, the earliest Chinese symbol (character) for paper denotes two cocoons hung on a thread, hinting either that silk fabric may have been the original material from which paper was invented or that paper replaced raw silk painted scrolls used as records. Yet, the thin, yellowish paper found by archeologists that dates but four years from the "invention" was made from vegetable fibers and not from silk.

Perhaps, as others suggest, Ts'ai Lun of the Eastern Han period (A.D. 25–220) merely perfected the art of making paper invented by his predecessors, the working people of the Western Han period (206 B.C.–A.D. 24). To the plant fibers used to make paper by his ancestors, Ts'ai Lun probably added macerated old fish nets, the bark of trees, and hemp, thus filling the needs of calligraphers and others for sheets of good quality and low cost on which to record events or practice their calligraphy.

There exists a mythical story about the invention of paper: Ts'ai Lun feigned death and arranged a false burial. His friends and followers were asked to burn paper money and paper charms over his grave to assure a happy and everlasting life for him in the future. (Ts'ai Lun, unknown to the mourners, was breathing fresh air through a bamboo tube.) When a sizeable amount of the paper turned to ash, Ts'ai Lun signaled his friends to exhume him, and thereby worked the first public relations gambit to become a "celebrity," initiated an instant custom, and assured widespread use of his invention.

Still others state that Ts'ai Lun recycled silk scraps by beating them to a fibrous pulp, using a stone

mortar and pestle. Then, after contemplating the problem further, he added this pulp, in small amounts, to a vat filled with water. Thus he had a vat filled with stuff. Perhaps he designed a mould, which may have been a cloth screen attached to a bamboo frame. He may have dipped the mould in the vat to catch the pounded cellulosic fibers on its surface and allowed the excess water to drain through the warp and weft of the cloth; or he may have unceremoniously dumped some pulp on to the mould and allowed the same thing to occur. In either instance, he probably placed the mould in the sun, allowed it to dry, and removed his magical piece of paper. Perhaps.

Certain accounts hint that Ts'ai Lun, in 114 A.D., was made a Marquis for his brilliant contribution to the court. Later, he was numbered among those caught up in a court intrigue opposed to a member of the Imperial family and, because he could not exonerate himself, committed suicide by taking poison.

Within 45 years, Tso Tzu-yi improved upon the methods and work of Ts'ai Lun; in a generation and a half, with regard to the practice of any art or craft (and I make no distinction between the two these days), it appears almost inevitable that men and women meliorate the processes of a given discipline.

The art of papermaking traveled both eastward and westward; it came to Japan via Korea, which was still a part of China. The Japanese behaved no differently toward Chinese manner, customs, arts, industries, agriculture, philosophy, science, literature, and religion than would individuals from any backwater province in any country toward the sophisticated minority in what was then regarded as the most highly developed culture in the world. Further, Buddhist missionaries from China "bringing culture to the masses" carried leaves and books of mulberry paper on

The Remarkable Journey of Handmade Paper from China to the New World

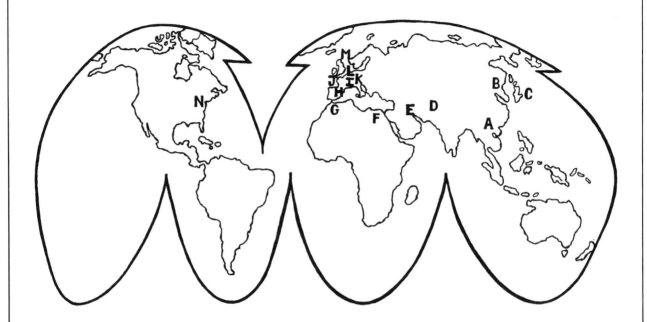

A. Ts'ai Lun credited with the invention of paper in Lei-yang, China, 105 A.D., although many authorities track the invention of paper to, at least, the second century B.C. The art is practiced in Tun-Huang in 150 A.D., in Loulan by 200 and in Niya about 250–300 A.D.

B. The invention reaches Japan via Korea in 610 A.D.

C. Empress Shotoku commissions "The Million Prayers," the first text printing on paper, in 770 A.D.

D. Chinese prisoners of war reveal secrets of paper manufacture to their Arab captors in Samarkand in 751 A.D.

E. The "secret" is carried to Baghdad, where Haroun-al-Rashid has paper manufactured in 793.

F. Egyptians make paper à la Chinois in 900.

G. Handmade paper appears in Morocco via Egypt in 1100.

H. Moors first introduce paper to Europe in Xativa, Spain (the present San Felipe near Valencia) c. 1151.

I. Oldest, uninterrupted paper mill built at Fabriano, Italy in the marquisate of Ancona in 1276.

J. First paper mill established at Troyes, France, c. 1348.

K. Ulman Stromer of Nuremberg sets up first mill in Germany with the assistance of Italian craftsmen in 1390.

L. Hollander machine invented in 1680, superseding all earlier devices including stamping mills and other primitive beating devices.

M. John Tate sets up a paper mill in Hertfordshire, England in the early sixteenth century, although the first mill established in England dates from about 1494. All paper, up to this time, was laid; John Baskerville, wanting a smoother wove paper for his printing in 1754, commissioned the Turkey Mill to produce a quantity for him.

N. Willem Ruddinghuysen van Mulheim (William Rittenhouse) creates the first paper mill in America on the banks of a small stream called Paper Mill Run, which flowed into Wissahickon Creek near Germantown, Pennsylvania in 1690.

and in which were recorded "wondrous strange" things. Thus, their curiosity piqued, students from Japan traveled to China to study and learn that which they deemed would be useful when they returned to their homeland.

A perplexing story suggests that one such Buddhist monk, by the name of Dokyo, brought the art and science of papermaking to Japan in the year 610 A.D. This physician, artist, ink-maker, and paper-maker must have been a dazzling long-lived fellow, for it is suggested that he became the chief physician and most trusted adviser of the Empress Shotoku, who reigned intermittently in Nara, the capital of Japan, from 749 to 769 A.D.!

Whether it was an act of penance for deposing and having had murdered her predecessor, the fear of another smallpox scourge similar to the epidemic of 735, or the guilt that weighed heavily on her royal head over the annihilation of those who participated in the rebellion of 764, the Empress Shotoku won her place in history as the person who invented and/or sanctioned the world's first text printing on paper of a *million* paper *dharani* (prayers), each enshrined in its individual wooden pagoda. This seven-year project was completed around 770 and is an indisputable and almost unbelievable fact.

Seventy years before this mass-media event occurred, sizing was introduced to papermaking. At first, gypsum was used to size, followed by a gelatine or glue, starch flour, and still other sizes obtained from various grains.

To follow the path of papermaking to the West, we are likewise forced to rely upon much speculation and some fact. In 707 A.D., we know that paper was used in Mecca. We can trace Marco Polo's route, and all of the caravans that preceded him, across the Gobi Desert and the equally arid Takla Makan, the Tarim Valley, and, finally, to Samarkand. From there, we can travel northeast to Turkestan and gaze upon the banks of

the Tharaz River where the Chinese fought the Arabs valiantly in 751. Another puzzling story suggests that a number of Chinese master papermakers were taken prisoner during this battle, and thus the "secret" of papermaking was "revealed" in Samarkand after 751 A.D. The story omits the fact that flax and hemp grew there in abundance, and that a complex network of irrigation ditches furnished the necessary water.

From Samarkand, the manufacture of handmade paper spread to Baghdad in the time of Haroun-al-Rashid; thus, the Arabs controlled two significant paper mills, both of which prospered considerably. A particularly white qualitative hempen paper was made in the Syrian cities of Bombyx, Damascus, and Tripoli, which perhaps explains the seeming reverence in the eyes of certain historians when they speak of *charta Bombycina* and *charta Damascena*—papers of supposedly extraordinary substance.

Twenty-five years after toilet paper was a common phenomenon in China, a thin paper, replacing papyrus, was manufactured in Egypt primarily for the purpose of copying hoary texts.

It is most curious that 1,000 years elapsed before paper left its inventor's hands in China to be manufactured in Europe. Why? Did the western world fear this inexpensive product which might bring learning to many? Was it fear of the Semitic peoples, who were the papermakers, introducing this new material as a substitute for their beloved parchment? Was it "secrecy" on the part of Moslem civilization to assure a fair profit in selling the end product—paper? Was the Mediterranean really a Moslem lake divided by a mile-high paper curtain? Was it any, or some, or all of these, or still other reasons in certain combination? As is said in the Southwest, "Quien sabe?"

Though certain scholars still dispute whether Italy or Spain may claim a "first" in the manufacture of

handmade paper, it seems that the edge is given to the paper mills of Xativa, Spain (the present San Felipe), near Valencia. These mills exported great quantities of paper to the West and the East from their origins in the mid-twelfth century.

The oldest, established, ongoing paper mill in the world is in Fabriano, Italy, near Ancona, where qualitative, handmade papers are still produced for persons who appreciate the best in the western world. The mill dates from about 1268 A.D.

Slowly, ever slowly, the manufacture of paper progressed from one European country to the next: to France in the fourteenth century; then four decades later, to Germany and to Flanders about 1405.

Why this snaillike pace? Primarily, paper, at this juncture, was not inexpensive. Its perishable nature made it a target for the sovereign contempt of the rulers and abbots, particularly those of Cluny. It was a product of Judeo-Arabic genesis, and that fact did not assist its growth, development, and usage in a Christian world that was over-zealous about anything that derived from Moslem acculturation.

Thanks to the fashion of wearing linen underwear instead of that itchy woolen stuff, there were cheap linen rags available to the mills, and the price of paper came down appreciably. From this point onward, paper manufacturers were dependent upon dealers in rags to produce their white art. Here is a fifteenth century warning from the city of Genoa: "Whoever does not belong to the said trade (of dealers in rags used in papermaking) is forbidden to sell used linen and old cordage; any one not belonging to that trade is forbidden to buy used linen and old cordage for resale in the city of Genoa, under penalty of a fine of four florins for each offense." (Blum, *On the Origin of Paper,* p. 37.) Eventually, after it had been manufactured throughout Europe, papermaking came to Mexico, the United States, and Canada.

Making a Papermaker's Hat

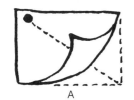

A

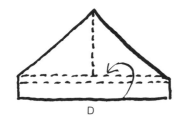

B

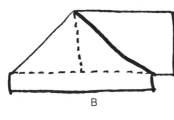

C

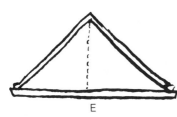

D

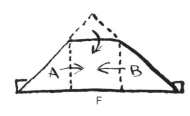

E

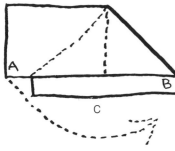

F

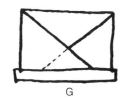

G

*T*radition or legend suggests that papermakers, because of cleanliness, a sense of union, ego fulfillment, or unknown reasons wore square, triangular, and other sorts of hats in addition to their green felt aprons (or other colors, depending upon custom in their country) when making paper.

Here is one way to make a papermaker's hat. If it does not fit properly, increase the dimensions of the original sheet proportionately or increase or decrease the top fold in G of the figure above.

1. Cut ½ a page from your local newspaper so you have a sheet of stock market quotations or your favorite advertisements 14½ × 18 inches in size.

2. With the 14½ inch side at your left, fold the paper in half by bringing down the top of the sheet. Now, fold it in half again by bringing the left side over to your right.

3. The double-folded sheet should be similar to A. Lift the top leaf and insert your index finger at the spot shown as you simultaneously bring the top sheet's lower corner down and left to form B.

4. Turn the sheet over (see C); hold your index finger at the bottom directly under the apex of the triangle. Then bring point A down and to the right to meet point B, creating a triangle. You will have to play with this step until the proper triangle is formed before smoothing down the folds. It should resemble D.

5. Fold the top sheet to the first dotted line; then fold it again to meet the base of the triangle.

6. Turn the piece over, and repeat the same step. Check your work against E.

7. Fold triangle A (the top one only) over to the right-hand side. A total triangle should still result.

8. Turn the piece over, and repeat step 7.

9. Between 2½ to 3 inches from the apex of the triangle, make a mark. Now, fold down the apex as shown in F. Fold over triangle A on the dotted line, as shown, and repeat for triangle B. Insert the rectangular ends one inside the other, and fasten with scotch tape, clips, or staples.

10. Turn the hat over, and repeat step 9. The hat should bear a decided resemblance to G.

11. Now, the moment of truth. Place your hands inside the hat, and push the folds outward to form H. Voilà!

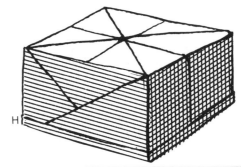

H

EARLY AMERICAN NOTABLES AND PAPER

Some of America's most famous statesmen played various roles in the development and promotion of paper: George Washington, for one, visited a paper mill run by Hendrick Onderdonk in what is now Roslyn, Long Island. When the vatman requested the President to try his hand and form a sheet, Mr. Washington obliged and even couched it. (There is no evidence of *this* aspect of the incident save for a note in the great man's diary, in which he mentions that he "breakfasted at a Mr. Onderdonk's at the head of a little bay where we were kindly received and well entertained. This gentleman works a grist and two paper mills, the last of which he seems to carry on with spirit and profit." (Hunter, *Papermaking Through Eighteen Centuries,* p. 257.)

Benjamin Franklin, for another, not only encouraged and offered financial support to 18 different mills, including the first one established in Virginia by his friend William Parks, but Franklin bought and sold large quantities himself. He was mostly interested in techniques—besides blowing his own horn even to the botanist Humphrey Marshall: "I was the more pleased to see in your letter the improvement of our paper, having had a principal share in establishing that manufacture among us many years ago, by the encouragement I gave it."

In addition to Washington and Franklin, Thomas Jefferson also had some association with papermaking. He wrote his first rough draft of the Declaration of Independence on Dutch paper, colored blue, about 8 × 12½ inches in size, bearing the watermark *Pro Patria Eiusque Liberate.* The lettering was composed in an oval border surrounding a lion and capped by a crown.

During the Civil War, rags were difficult to locate at any price, and it seems that sundry American businessmen startled their competitors by importing Egyptian mummies, thereby solving *their* problem with millions of yards of linen wrappings. (They simultaneously *provided* the problem of cholera to the rag pickers and rag cutters of their time.) The paper produced from the mummies was supplied to butchers for wrapping meat!

ON INTERNATIONAL PAPER SIZES

Theory and practice with regard to a universally accepted set of sizes (lengths, widths, and weights) appear to be at variance with one another, depending upon the country and time. Handmade paper dealers, handmade paper manufacturers, and artists who use fine papers all seem to hold conflicting views.

In 1943, Dard Hunter wrote of some 240 different sizes of European printing or writing papers then being manufactured including: Antiquarian, 31 × 53 inches; Atlas, 26 × 34 inches; Billet Note, 6 × 8 inches; Colombier, 23½ × 34½ inches; Crown, 15 × 20 inches; Demy, 17½ × 22½ inches; Elephant, 23 × 28 inches; Double Elephant, 26¾ × 40 inches; Emperor, 48 × 72 inches; Foolscap, 13½ × 17 inches; Grand Eagle, 28¾ × 42 inches; Hand, Middle, 16 × 22 inches; Hand, Royal, 20 × 25 inches; Imperial, 22 × 30 inches; Post, 15½ × 19½ inches; Pott, 12½ × 15½ inches; and Royal, 20 × 25 inches.

In 1818 England, punishment was exacted if a manufacturer made a newspaper larger than 22 × 32 inches!

It seems either premature, an example of wishful thinking, or evidence of supreme provincialism, but a list of international paper sizes is in existence. Perhaps soon it will become widely used.

CRYSTAL BALL GAZING

In addition to the biologists, chemists, engineers, foresters, horticulturists, physicists, super-specialists in the spaces between these areas, technicians, and armies of skilled hands engaged in continuing research in papermaking for industry, artists, artist-craftsmen, and creative persons are intuitively attacking new and unusual ways of producing papers to meet their personal needs. The latter, if they have reached the age of wisdom—whenever that is supposed to happen—may try to adapt what scientists have evolved; may reach out to make quantum leaps that carry them beyond the usual practice; may find in the ordinary, the pedestrian, the everyday garden-variety phenomenon about us, a new way of approaching the old problem of how to make paper.

Paper is now being made from refuse and garbage; waste paper is being recycled, after undergoing shredding and de-inking; synthetic materials are being developed to substitute for pulp derived from wood; Finnish and Danish engineers produce paper without water, resulting in sheets comparable to cotton, linen, silk, or velvetlike finishes, it is reported; Dutch engineers are working on producing papers through electrostatic means.

New, faster, more economical machines are being invented to replace the Fourdriniers including the Multi-Former, Periformer, Ultra-former, and the Verti-Forma. (The latter, as implied by its trade name, makes paper in a vertical plane as opposed to the long, continuous horizontal plane of the Fourdrinier.)

There is little doubt in anyone's mind that we stand at the crossroads to an unusual future with regard to the making of paper. Wherever, however, whatever, whenever, you look backwards from the unknown future, it is deeply hoped the "Edward Bellamy glance" proves useful.

A Concise Biased Chronology

	B.C.	
Indus Valley sites flourish in West Pakistan	3000	
Sphinx built in Egypt	c.2900	
	2700	Ts'ang Chieh conceives Chinese "characters." (Calligraphy invented.)
Hsia Dynasty—China	c.2000 –c.1500	
Babylon ruled by Hammurabi	c.1792 –1750	
End of Mohenjo Daro and Harappa civilization in West Pakistan	1500	
Israelites led out of Egypt by Moses	c.1450 or c. 1275	
	1400 –1300	Chinese incise divination bone writing
Ikhnaton rules Egypt	1360	
Fall of Troy	1184	
	900	Greeks manufacture felt
Romulus founds Rome	753	
Babylonians destroy Ninevah	612	
Birth of Gautama Buddha (d. 483) in Nepal	563	
Pythagoras dies	500	Chinese scholars *still* write on bamboo strips
Confucius born (d. 478)	551	
Athenians defeat Persians at Marathon	490	
Thermopylae Pass— Persians defeat Greeks	480	
Athenians rout Persians at Plataea	479	
Peloponnesian Wars begin Sparta vs. Athens	431	
Plato born (d. 347)	429	
	400	Silk widely used for books and calligraphy
Socrates a suicide by hemlock	399	
Alexander the Great born	356	
Mayan calendar invented	300	
First Punic War: Rome vs. Carthage	264	
	255	Seals impressed into clay
	250	Meng t'ien invents camel's hair brush
Second Punic War: Hannibal crosses Alps; defeats Romans	218 –146	

	200	Parchment manufacture refined in Pergamum
Third Punic War: Carthage destroyed	149 –146	
Julius Caesar assassinated	44	
Christ born in Bethlehem	4	

	A.D.	
Jesus crucified	29	
Christians persecuted by Nero	64	
Jerusalem destroyed by Titus	70	
Mt. Vesuvius erupts	79	
	105	Ts'ai Lun "invents" paper
	150	Tso Tzu-yi improves the process
Council of Nicaea	325	
Rome sacked by Goths	410	
Justinian born (d. 565)	483	
	500	Mayans already using *amatl* (bark paper)
Aesop dies	550	
Mohammed born (d. 632)	570	
	610	Paper used in Japan
Arabs conquering North Africa	670	
	700	Sizing introduced
	707	Paper used in Mecca
Golden Age of Mayans begins	731	
Charles Martel defeats Moors	732	
	770	Empress Shotoku's million *dharani* (prayers) printed
	794	Paper made in Baghdad
Charlemagne named Holy Roman Emperor on Christmas Day	800	Paper used in Egypt
	807	Paper made in Kyoto, Japan
	875	Toilet paper used in China
King Alfred dies	899	
	900	Paper made in Egypt
	950	Paper used in Spain
	969	First mention of playing cards, China
Leif Ericssons' voyage to Finland	1000	
	1035	Paper-wrapped vegetables in Egypt

Event (left)	Year	Event (right)
	1041–9	Movable type invented by Pi Sheng, China
Battle of Hastings; England conquered by William of Normandy	1066	
First of the many Crusades	1096	
	1100	Paper used in Istanbul
	1102	Paper used in Sicily
	1150	Paper made in Xativa, Spain
	1154	Earliest watermark?
Genghis Khan born (d.1227)	1162	
	1189	Paper used in France
Magna Carta	1215	
	1228	Paper used in Germany
	1250	Block printing in Egypt
	1268	Paper made at Fabriano (still going strong)
Marco Polo off for Cathay	1271	
Thomas Aquinas dies	1274	
	1282	Watermarks in wide use
	1285	Fleur de Lys watermark used
Dante and Giotto shine forth	1300	
	1309	Paper used in England
Giovanni Boccaccio born (d. 1374); Gunpowder invented	1313	
	1322	Paper used in Holland
Hundred Year's War begins	1337	Animal sizing first used in Europe?
Bubonic plague reaches Venice	1348	Paper made in France
	1390	Paper made in Germany
	1403	Movable type produced in Korea
	1423	Block printing in Europe
	1428	Paper made in Holland
Joan of Arc burned to death at Rouen by the English	1431	
	1433	Paper made in Switzerland
100 Year's War ends (England vs. France); Turks capture Istanbul	1453	
	1456	Johann Gutenberg completes first Bible from movable type
	1465	Blotting paper in use
Balboa, Spanish explorer born (d. 1519)	1475	

Event (left)	Year	Event (right)
	1476	William Caxton of England sets up first printing shop; 30 books in first three years
	1479	Fool's Cap watermark used
	1491	Paper made in Poland
Christopher Columbus sails west	1492	
	1494	Paper made in England
	1495	First watermark in England
John Cabot reaches Canada, Amerigo Vespucci disputes Columbus' claim	1497	
Vasco da Gama reaches India from western Europe; Savonarola burned as heretic in Florence	1498	
	1506	Bramante, Michelangelo, and Raphael commissioned by Pope Julius II to paint in St. Peter's
Martin Luther posts his 95 theses	1517	
Hernando Cortes begins conquest of Mexico	1519	
Fernando Magellan's crew circumnavigates the world	1522	
Giovanni da Verrazano explores New England coast for French; New York bay?	1524	
Francisco Pizarro conquers Peru	1531 –35	
John Calvin publishes his Institutes	1534	
Jacques Cartier discovers St. Lawrence River	1534 –36	
	1540	Glazing hammer introduced in Germany
Hernando de Soto discovers Mississippi River	1541	
Council of Trent	1545	
	1550	Smalts used to color paper blue
William Shakespeare born (d. 1616); also Galileo born (d. 1642)	1564	
	1575 –80	Paper made in Culhuacan, Mexico
	1576	Paper made in Russia
Mary, Queen of Scots executed for treason by Elizabeth I	1587	

Historical Event	Year	Papermaking Event
Spanish Armada destroyed	1588	
	1591	Paper made in Scotland
Guy Fawkes Gunpowder Plot foiled	1605	
Rembrandt born (d. 1669)	1606	
Capt. John Smith lands in Virginia	1607	
Henry Hudson sails from New York harbor to Albany; Spaniards settle Santa Fe, New Mexico	1609	
King James version of Bible printed	1611	
Mayflower lands at Plymouth Rock, Cape Cod	1620	
Harvard College founded; plague brought to London through papermaker's rags	1636	
Taj Mahal completed	1648	
Great Plague of London	1665	
	1680	Hollander beater "perfected" in the Netherlands
Johann Sebastian Bach born (d. 1750)	1685	
	1690	Paper made in Philadelphia by William Rittenhouse
Benjamin Franklin born	1706	
	1710	William De Wees sets up second paper mill near Philadelphia
Slaves revolt in New York (also in 1741)	1712	
	1726 –28	William Bradford makes paper in New Jersey
	1728	Paper made in Massachusetts, near Boston
	1733	China clay discovered by William Cookworthy, England
	1734	Paper made in Maine
Joseph Michel Montgolfier born (d. 1810) —is papermaker who later will make the first balloon flight	1740	French standardize sheet size and watermarks; Isaac Langle, makes paper moulds, Philadelphia
	1744	Paper made in Virginia
Goya born (d. 1828)	1746	
	1755	Wove paper made!
Formal declaration of French and Indian War (Seven Year's War)	1756	
	1757	Death of Réamur who, in 1719, first suggested wood as a papermaking fiber
	1758	First forgery of an English bank note; 2nd Volume of Robert Dossie, *Handmaid to the Arts* published—contains a chapter on *papier maché* use for decorative ceilings and bas-reliefs in homes
	1760	Watermarks used in wove paper
	1767	Paper made in Connecticut
Napoleon Bonaparte born (d. 1821)	1769 –73	Paper made in New York
Ludwig Van Beethoven born (d. 1827)	1770	Nathan Sellers makes paper moulds, Philadelphia
	1772	Dr. Christian Schaffer completes six-volume treatise on vegetable fibers for papermaking; in Europe, paper used for building coaches, cabinets, screens, bookcases
	1773	Death penalty for forging watermarks in English bank notes
	1774	Chlorine discovered independently by K.W. Scheele of Sweden and by C.L. Berthollet of France in 1780s—used to bleach paper stocks
American Revolution starts	1776	Earliest paper mill in central Massachusetts
	1777	Paper made in North Carolina; Charles Tinnant develops bleaching powder
End of American Revolution	1783	
Shay's Rebellion; Northwest Ordinance adopted; Constitutional convention opened	1787	Paper made in Delaware by Gilpin family
British penal colony established in Australia	1788	Ben Franklin describes how to make large sheets of paper in the "Chinese manner"
George Washington chosen president; French Revolution begins	1789	
Benjamin Franklin dies	1790	Hydraulic press invented by Joseph Bramah, England
	1790 –95	Paper made in Vermont
	1792	Paper made in New Hampshire
	1793	Paper made in Kentucky

Event	Year	Paper/Technology Event
The Whiskey Rebellion in Pennsylvania	1794	
Washington's Farewell Address	1796	
Washington dies	1799	N.L. Robert patents a paper machine
	1802	"Hog" invented in England; paper made in Quebec for *Montreal Gazette*
Louisiana Purchase	1803	
Code Napoleon adopted in regime of Napoleon Bonaparte; Lewis & Clark Expedition; Burr-Hamilton duel	1804	First book printed on machine-made
Napoleon defeats Prussians at Jena	1806	Henry Fourdrinier granted patent for paper machine; John Pine establishes the Hayle Mill, Tovil, Maidstone, Kent, England (acquired by the Green family in 1811 and still operating); paper made in South Carolina
Robert Fulton's steamboat; Aaron Burr tried for treason	1807	Paper made in Ohio
	1810	Paper made in Georgia
	1811?	Paper made in Tennessee
Napoleon invades Russia	1812	
Louis XVIII restored to French throne; Napoleon exiled to Elba	1814	
	1817	First paper machine in America erected in Thomas Gilpin's mill near Philadelphia
	1819	Sir William Congreve patents colored watermarks; second paper mill established near Halifax, Lower Canada
Monroe Doctrine declared; Samuel Brown of London successfully operates an an internal combustion engine	1823	Gypsum (calcium sulphate) used as a loading material for paper in Europe
Erie Canal opened; first steam locomotive in U.S. built by John Stevens of Hoboken, New Jersey	1825	Dandy roll invented; James Crooks establishes a paper mill in Lower Canada
	1826	Paper made in Indiana
	1834	Paper made in Missouri and Michigan
	1839	Joynson patents machine-made watermarks
Invention of the telegraph	1837	
Antarctic found to be continent	1840	Paper made in Illinois; watermarks in first postage stamps
First covered wagon train for California left from Independence, Missouri; first passenger train on Erie Railroad	1841	Charles Fenerty of Nova Scotia makes first groundwood paper in western hemisphere (on a laboratory basis)
First telegraphic message by inventor, Samuel F.B. Morse; "What hath God wrought?"	1844	
	1845	W.H. Smith of England invents shadow marks
War between U.S. and Mexico	1846	
Adhesive U.S. postage stamps used	1847	
Marx's *Communist Manifesto;* gold discovered in California; Louis Philippe dethroned in France	1848	Paper made in Wisconsin
	1849	Paper made in Alabama
Walden by Henry Thoreau	1854	Paper made in Utah
Leaves of Grass by Walt Whitman; *Song of Hiawatha* by Henry W. Longfellow	1855	Egyptian mummy wrappings used in U.S. to make paper for grocers, butchers, etc.
	1856–57	Paper made in California
Dred Scott Decision; Great Mutiny in India (Sepoy Rebellion)	1857	
Harper's Ferry seized; *Origin of Species* by Charles Darwin	1859	Paper made in Minnesota
Abraham Lincoln elected President of U.S.; Pony Express begins; Garibaldi unifies Italy	1860	Joseph Jordan invents a cone-type refiner (continuous beater)
Abraham Lincoln assassinated	1865	
Ku Klux Klan organized	1866	Paper made in Oregon and Iowa; Alexander Buntin establishes first mechanical pulp mill in Canada
Transcontinental railroad completed; Black Friday in New York financial circles	1869	Shadow-marked dandy invented
Franco-Prussian War ends; Mrs. O'Leary's cow and the Great Chicago Fire; Wilhelm I now Kaiser of Germany; "Dr. Livingstone I presume?"; Emily Carr, Canadian artist born (d. 1945)	1871	Seth Wheeler patents roll-form toilet tissue
W.C. Handy born (d. 1958)	1873	
"Boss" Tweed convicted of fraud	1874	Paper made in Kansas
President James Garfield assassinated; Picasso born (d. 1973)	1881	Paper made in Nebraska

Event	Year	Papermaking event
Brooklyn Bridge opened	1883	
	1884	J. Hoyt invents his beater
Louis Riel hanged for treason in Canada; last spike driven in Canadian Pacific Railroad; Ferdinand "Jelly Roll" Morton born (d. 1941)	1885	Paper made in Washington; William Umpherston of Scotland patents his beater in U.S. (also 1884)
Haymarket Riot; Sherlock Holmes created by Dr. Arthur Conan Doyle	1886	
First electrocution, Auburn, New York; Battle of Wounded Knee	1890	Toilet paper beginning to be used in U.S.
Serge Prokofiev born (d. 1953)	1891	Paper made in Colorado
X-rays discovered by W.K. Roentgen	1895	Mould-made paper machine introduced in England
Radium discovered by the Curies and G. Bemont in Paris; George Gershwin born (d. 1937)	1898	Paper made in Louisiana
Boxer Rebellion; Carrie Nation wields her ax; yellow fever fought; Louis Armstrong born (d. 1971)	1900	Paper made in Florida
President William McKinley assassinated	1901	English beginning to use paper for oil drums, hansom cabs, drain pipes
Wright brothers' maiden flight	1903	First use of fiber boxes in U.S.
	1905	Glassine paper introduced to the U.S.
San Francisco earthquake and fire	1906	First paper milk bottles used in San Francisco
Wall Street panic over U.S. financial condition	1907	Brompton Paper Mills in Quebec made the first kraft (sulphate) pulp in North America
Peary reaches North Pole; Bleriot flies the English Channel	1909	First kraft paper made in U.S.
	1910	Bread and fruit wrapped in special papers in U.S.
Alexander Graham Bell and Thomas A. Watson hold first phone talk New York to San Francisco	1915	Paper trays used for raisin drying in California (formerly, wood employed)
	1919	Eden Philpotts publishes *Storm in a Teacup*, first accurate novel written in English on handmade papermaking
First meeting of the League of Nations at Geneva; 19th Amendment to Constitution adopted (women's suffrage)	1920	A Wisconsin paper mill on a Fourdrinier machine made paper at a rate of 1,000 feet per minute
	1921	*Birmingham Age-Herald* of Alabama printed entirely from Alabama spruce pine for first time

Event	Year	Papermaking event
Nicola Sacco and Bartolomeo Vanzetti executed for purported murder of two men in payroll robbery	1927	Claude Dravaine writes novel on old French papermaking—not as technical as Philpotts (see 1919)
First talking picture, "Lights of New York"; Kellogg-Briand Peace Pact signed by 62 nations condemning war	1928	Dard Hunter revives handmade papermaking in Lime Rock, Connecticut mill
	1931	Dard Hunter's mill closes down
The Dionne quintuplets born in Canada	1934	At Iwano Mill, Okamoto, Echizen, Japan makes the then largest handmade sheet of paper: 200 × 200 inches (40,000 square inches), made by spraying over a porous surface laid above the floor
First nuclear chain reaction at University of Chicago under Enrico Fermi and Arthur Compton	1942	Estimation that about 100 tons of paper are required in the course of building a battleship of the Massachusetts class—16 tons of blueprints; the remainder letterheads, contracts, graphs, stencils, etc.
President Roosevelt dies; Mussolini executed by partisans; atomic bomb dropped on Hiroshima and Nagasaki; Hitler and wife, Goebbels and family suicides; Quisling executed; U.S. Forces enter Korea	1945	Someone estimates there are approximately 14,000 different products made from paper
Nuremberg War Crimes trials; Philippines granted independence	1946	
Fidel Castro leads Cuban revolution; St. Lawrence Seaway opens; Premier Khrushchev visits U.S.	1959	
Bay of Pigs; Dag Hammarskjold killed in plane crash; Yuri Gagarin first human in orbit	1961	
John F. Kennedy assassinated	1963	
6-day Israeli-Arab War; Dr. Christian Barnard performs first human heart transplant	1967	
The Reverend Dr. Martin Luther King assassinated; Senator Robert F. Kennedy killed; U.S.S. Pueblo seized by North Korea. Charles de Gaulle resigns; Neil Armstrong first man on the moon	1968	
Voyagers I and II depart earth for deep penetration of our solar systems	1977	

METRIC CHART

METRIC CONVERSION FACTORS (Approximate)

Symbol	When You Know Number of	Multiply By	To Find Number of	Symbol
LENGTH				
in	inches	2.54	centimeters	cm
ft	feet	30	centimeters	cm
yd	yards	0.9	meters	m
mi	miles	1.6	kilometers	km
AREA				
in^2	square inches	6.5	square centimeters	cm^2
ft^2	square feet	0.09	square meters	m^2
yd^2	square yards	0.8	square meters	m^2
mi^2	square miles	2.6	square kilometers	km^2
	acres	0.4	hectares	ha
WEIGHT (mass)				
oz	ounces	28	grams	g
lb	pounds	0.45	kilograms	kg
	short tons (2000 pounds)	0.9	metric tons	t
VOLUME				
tsp	teaspoons	5	milliliters	mL
Tbsp	tablespoons	15	milliliters	mL
in^3	cubic inches	16	milliliters	mL
fl oz	fluid ounces	30	milliliters	mL
c	cups	0.24	liters	L
pt	pints	0.47	liters	L
qt	quarts	0.95	liters	L
gal	gallons	3.8	liters	L
ft^3	cubic feet	0.03	cubic meters	m^3
yd^3	cubic yards	0.76	cubic meters	m^3
PRESSURE				
inHg	inches of mercury	3.4	kilopascals	kPa
psi	pounds per square inch	6.9	kilopascals	kPa
TEMPERATURE (exact)				
F	degrees Fahrenheit	5/9 (after subtracting 32)	degrees Celsius	°C

GLOSSARY

ABACA. Manila hemp; also the name of the plant.

ABRASIVENESS. That property of a paper surface which may scratch or cut; undesirable for printing.

ABSORBENCY. The capacity of paper to absorb liquids.

ALPHA CELLULOSE. That part of a cellulosic material that fights solution by aqueous caustic alkalies at ordinary temperatures.

ALPHA PROTEIN. Soybean protein that used alone or with casein, makes an adhesive with which to coat paper.

ALUM. Aluminum sulfate. It is used in the beater to precipitate rosin sizing on to the pulp and provides water resistant properties to the paper.

ANILINE DYES. Derived from coal tar, they are arranged with regard to their brightness or fastness to light.

APPARENT DENSITY. The weight of a sheet of paper determined by dividing the basis weight by the caliper or thickness.

ASH. That which is left after burning a sample of paper to determine the amount of filler it contains.

ASP. See Ass.

ASS. A notched piece of wood which fits into the assboard to hold the mould at a particular draining angle. Also called Asp or Horn.

BASIS WEIGHT. See LBS and g/m²

BEATING. Process of macerating natural or manmade material into pulp.

BEDPLATE. The bars or knives of the fixed plate in the floor of a Hollander-type beater directly under the beater roll.

BLANC FIXE. Artificial barium sulphate or precipitated barium sulphate.

BLEACH. Chlorine, or a similar chemical, used by the paper industry to whiten paper pulp.

BLEEDING. That phenomenon which occurs when the edges of a color begin to dissolve as a result of water or oil on the paper.

BLISTER. A paper defect that may be caused by any number of reasons including trapped air between the felt and the sheet.

BONDING STRENGTH. The property of a sheet that allows it to withstand "picking," the pulling away of part of the surface, especially in large, solid-printed areas. Of prime interest to printmakers.

BRIDGE. A platform across the vat.

BRIGHTNESS. Once associated with the hue and intensity of light reflected from the paper surface to the eye. Now scientifically measured by instruments.

BRITTLENESS. A paper defect evidenced when paper breaks or fails when bent.

BROKE. Waste pulp and damaged paper.

C1S. Coated on one side of paper.

C2S. Coated on both sides of the paper.

CALCIUM CARBONATE. A solid appearing in nature as calcite or aragonite and used because it is white to coat paper.

CALENDER BLACKENING. If the moisture of the paper is too high, dark streaks or dark areas result when trying to calender the paper.

CALENDERING. The passing of paper between metal rolls (as in an etching press) or the placing of paper between metal plates and further squeezed to increase smoothness or gloss of paper.

CALIPER. Thickness of a sheet of paper, normally measured in thousandths of inches.

CANVAS. A triple cloth used as the base upon which large sizes of handmade papers are dried in lofts. Used to be made of Hessian trebles. Today, many substitutes are used.

CASEIN. A protein obtained from skimmed milk used as an adhesive in paper coating.

CELLULOSE. An inert substance constituting the chief part of the cell walls of plant materials, trees, paper, and so on.

CHINA CLAY. A fine clay used, while beating, as a filler for certain papers.

CLOSE FORMATION. A uniformly dense sheet of paper. Opposite of cloudy.

CLOUDY. A "wild" or flocculated sheet of paper. Unevenly formed, when held held up to the light.

COATING SLIP. A slurry of pigment and adhesive used to coat papers.

COCKLING. A wavy effect caused by uneven drying.

COLLOIDAL. Any substance in a very fine state of dispersion.

COMPRESSIBILITY. That property of a sheet which allows it to withstand pressure, such as in printing.

CONTRARIES. Unwanted, unneeded bits of materials which become parts of the sheet.

CORNER UP. Dog-eared corners of sheets.

COUCH. The transfer of the layer of wet pulp on a mould to a dampened felt.

COW HAIRLINE. Lines on which paper was hung to dry in the drying loft. They were made of jute and wound around with cow hair to prevent staining on the backs of the sheets.

C.P. Cold-pressed paper: a polished, glossy paper with a hard surface. Obtained by sandwiching papers between zinc sheets in a wringer-type press.

CRACKING. Visible breaks along creases of sheets when paper is folded or embossed.

CRINKLED. Either a deliberate crèpe effect in industry or a defect in handmade paper due to movement at the time of couching.

CROSS. Resembles a T-square. Used, in the past, to handle papers for placing or removing from the cow hairlines in the drying loft.

CURL. Rolled-up edges caused by changes in temperature and humidity.

DANDY ROLL. In the paper industry, a large cylindrical roll on the "wet end" of the Fourdrinier, which impresses a watermark on the paper. This watermark, when the paper is held to the light, is darker than the ground around it. Opposite effect of raised wire watermark on a mould.

DECKLE. The removable frame that fits snugly on the mould and contains the pulp.

DECKLE EDGE. The feathered edge on all four sides of a sheet of handmade.

DECKLE SLIP. A slip of brass affixed to the underside of the deckle to prevent the pulp from getting under the deckle. Used with a new mould, especially.

DISHED. Refers to a stack of paper that lies in a decidedly concave condition rather than flat.

DROPS. Drops of water that truly make watermarks on newly formed sheets, thus creating a "defect" which may be seen by holding the sheet to the light. Come from the arms of the coucher or vatman, if not well-trained.

DRYING. Usually refers to loft drying of waterleaf, though applicable to all papers including those just sized.

DRY PRESSING. The final pressing given a particular sheet of handmade paper before it leaves the press area and is counted and wrapped.

DRYWORKERS. Traditional term for those individuals who assist in drying paper in the lofts of paper mills.

EMBOSSED. A noninked intaglio or relief design imparted to paper.

EXPANSION. The result of change in the dimensions of a sheet of paper due to excess humidity; worse across the grain, rather than with it.

FASTNESS. That evanescent quality sought for by all papermakers: resistance to change in color of dye used; paper should also be fast to acid and alkali.

FELT. A woven blanket on which paper is couched from the mould to the felt; also, to remove water from a sheet with a sponge (using the "easiest method" of papermaking).

FELT MARK. A defect in handmade paper caused by a worn felt.

FIBRILLAE. One of the effects of the Hollander upon cellulose fibers —separation of the threadlike elements on each fiber.

FIBRILLATION. The shredding and bruising of fibers by the beater bars in the Hollander.

FILLER. One method of filling in the pores of a sheet to improve its printing qualities. See also Loading.

FINISHING. Drying, sizing, and calendering papers to complete the manufacture of a given sheet or sheets.

FISH EYES. Defects in paper caused by foreign matter, slime, etc., which become translucent spots when the sheet is calendered.

FOLIO. Either a sheet of paper 17 x 22 inches in size, or a sheet which is folded in two.

FOURDRINIER. The term usually given the ubiquitous paper machine. It should apply solely to the wet end of the machine. The original instrument was designed by N.L. Robert and financed by the Fourdrinier brothers.

FREE. Pulp that drains quite readily from the papermaker's mould.

FUR. A defect caused by stuff, through improper couching, adhering to a previous felt and transferring itself to a newly formed sheet.

FURNISH. The particular ingredients that comprise a specific paper. One "furnishes a beater" with the items in a given formula.

FUZZ. Fibrous projection on a paper surface caused by lack of surface sizing, or insufficient beating. Lint may seem similar, save that it is not locked to the surface.

GELATINE. One of many sizings that may be used to make waterleaf papers less liable to bleed; a glutinous material obtained from animal tissues through continued boiling.

GLAZE BOARDS. Thick manila boards used to obtain a matte, eggshell finish on certain papers by running a "sandwich" of paper and boards through calender rolls.

GLAZING. Process in which a pile of sheets, between each of which lies a zinc plate, is submitted to the pressure of steel rollers causing a slight friction which glazes the sheets

g/m². Grams per square meter (a preferred mode of expressing weight of paper). See LBS.

GRAINY. A rough finish on the surface caused by shrinkage of the sheet, under certain conditions.

HALF-STUFF. Half-beaten stock purchased from a paper mill. It must be subjected to further beating in a hollander before proper pulp results.

HAND BASIN. A basin, secured to the vat, which allows the vatman to cleanse his hands of knots of pulp from time to time.

HESSIAN TREBLES. See Canvas.

HOG. A device for constantly stirring the pulp in the vat.

HOLLANDER. A seventeenth century-designed beater still employed by handmade papermakers to bruise, or macerate, rags to fibers.

HORN. See Ass.

H.P. Hot-pressed paper; a polished, glossy paper with a hard surface. Used to be calendered between hot metal sheets passed through a wringer-type press.

HYDRATION. The process that, through beating, alters cellulose fibers so as to increase their water absorption capabilities.

HYGROEXPANSIVITY. Expansion or contraction of paper due to changing conditions of humidity.

HYGROMETER. An instrument which measures the relative humidity of air in a given place.

HYGROSCOPIC. Water-loving; easily absorbs moisture.

INK ABSORPTION. That property of a sheet that allows it to "take" ink quickly (as in newsprint), or the reverse.

JORDAN. A machine which refines pulp after it has gone through a beater. Used in the paper industry before pulp travels to the Fourdriniers.

JUTE. Very strong, long-fibered pulp made from hemp, used burlap, and twine.

KAOLIN. White, fine clay used in some formulas for papermaking.

KNOTTER. An agitated strainer containing certain sized holes (for particular papers) that keeps out knots when the pulp is strained on its way to the vat.

LAMINATED PAPER. Two or more sheets couched one upon the other.

LAYER. In the trinity of vatman, coucher, and layer, the latter—after the paper is pressed for the first time—removes the felts carefully and lays each sheet of paper on a zinc plate covered with a felt and, one by one, creates another pack and another. . . .

LBS. The weight of a 500 sheet ream of paper of specified dimensions. The higher the number, the heavier the paper. Also called Basis Weight.

LINTERS. Usually cotton. That short, cotton fiber that remains after the ginning operation. Usually comes in three grades: first cut, mill run, and second cut.

LOFT. A large, airy room (capable of being heated) in which handmade has traditionally been dried.

LOFT TEMPERATURE. In loft drying, it has been considered a truism that papers not be submitted to temperatures above 90° F (32 degrees C.).

LOUVRES. Shutters in old paper mills in the loft drying areas that allow proper air circulation.

MACERATE. To bruise and separate individual fibers of natural or manmade material by beating by hand or machine.

MATURING. Aging of paper. Ideally, paper should age considerably before being used by painter, printmaker, or printer. Some idealists would like to wait 300 years!

MELAMINE RESIN. A chemical added to the stuff in the beater when wet-strength is desired in the paper.

MOISTURE CONTENT. A figure that

varies from day to day: the percentage of moisture in finished paper.

MOULD, LAID. A mould made of vertical (laid) heavy wires bound by chain wires and supported by tapered, wooden ribs in a wooden frame.

MOULD, WOVE. A tautly stretched, nonferrous woven wire screen across a wooden frame supported by tapered wooden ribs.

OUTSIDES. Paper with very bad faults. Usually repulped. Literally used (1 quire each) at the top and bottom of reams, at one time.

PACK FELTS. A zinc plate covered with felt. See Layer.

PACK PARTING. Separating pressed sheets from the zinc plates.

PACK PRESSING. Submitting a pack of dampened felts interleaved with newly formed sheets to tremendous pressure, usually, in a hydraulic press.

pH VALUE. On a scale from 0 to 14 (0–7 is acid; from 7–14 is alkaline), the degree of one or the other may be obtained.

PICKING. The lifting, during printing, of portions of the paper surface— especially in large black or color areas.

PILCHER. The term given to the three or four felts, sewn together, and placed on top of a newly formed post of paper before it is taken to the press.

PINHOLES. Defects in sheets that appear, when held to the light, as pinholes, Caused by foreign particles pushing through the sheet when it is being pressed.

POST. A pile of wet, handmade sheets separated by dampened felts.

PRINTING OPACITY. That property of a sheet which prevents the printing on one side from showing through on the other.

PULP. That fibrous substance resulting from the pulping process which still requires further beating before it is usable for forming paper.

QUIRE. Usually, 24 sheets. One-twentieth of a ream.

RATTLE. The sound, made by shaking a sheet of paper, that indicates its rigidity, stiffness, or lack of dampness.

REAM. Usually 20 quires. Thus, there may be either 480 or 500 sheets in a given ream, depending upon the size of the sheets and/or practice in the given country.

REAM WRAPERS. The final act before a ream of handmade paper left the papermill was to affix the ream wrapper to the pack. Paper historians and collectors seek them.

REFINE. Used by some individuals instead of the word, beat.

RESILIANCE. The property of a sheet that allows it to return to its original format after distortion.

RETREE. Handmade sheets containing minor defects.

ROSIN. Used in sizing paper. Derives from the distillation of turpentine from the gum of the southern pine.

SHAKE. The vatman's stroke. A highly individualized movement from right to left, left to right, and from the body away to the vat. Takes strength, grace, and many years to master.

SHEARING STRENGTH. That property of a sheet which allows it to resist cutting.

SIZING. A water-resistant material added to paper.

SLURRY. Stock to which the proper amount of water is added to form a suspension in which you can form paper with a mould and deckle.

SNOWSTORM. A term used by some to indicate a "wild," flocculated paper.

SPECIFIC VOLUME. Assuming standard conditions, that volume per unit mass measured.

SPUR. A particular number of sheets picked from a pack of paper and either hung to dry or placed on Hessian trebles to dry.

STANDARD SIZES. Despite so-called international sizes of papers, there does not appear to be uniform practice.

STAY. A board across the vat on which the vatman first places his mould after forming a sheet.

STOCK. Pulp ready to be formed into sheets.

STUFF. Untreated papermaking pulp.

SULFATE. In industry, the alkaline or kraft process of brewing pulp.

SULFITE. In industry, the acid process of brewing pulp; the pulp produced by this process.

TENSILE STRENGTH. The property of a sheet to withstand pulling.

TEXTURE. A difficult-to-describe "feel" of the surface of a sheet to your touch.

THICK EDGES. A couching defect caused by uneven pressure at one end during the forming of the sheet; the pulp end may double up on itself and form a thick edge.

TITANIUM DIOXIDE. A filler used in papermaking to brighten and opacify certain sheets.

TOOTH. A description of the surface of a sheet of paper implying a slightly rough texture having, per-haps, too many "hills and valleys."

TRANSLUCENCY. That property of a sheet which allows light to pass through to a certain degree. *Not* transparent.

TRANSPARENT. That property of a sheet which allows you to see and to distinguish objects *through* a sheet.

TREBLES. See Canvas or Hessian Trebles.

TUB-SIZED. Paper sized in a tub or vat after it is formed as opposed to paper sized by adding the size to the beater.

VATMAN. The keystone in the trinity of individuals who make handmade papers. On the vatman's skill and stroke rests the quality of the operation. See also Coucher and Layer.

VAT. A rustproof, waterproof tub holding a sufficient amount of pulp with which to make paper.

WATER ABSORPTION. That property of a sheet that allows or discourages water to be soaked up or repelled.

WATERLEAF. Unsized paper.

WATERMARK. The logo or identifying mark of the papermaker or mill at which a sheet of handmade paper was formed. It may be seen by holding the paper up to the light.

WAVY EDGES. A paper problem that occurs in a pack when the edges of the sheets have acquired moisture and expanded.

WILD. See Snowstorm.

XX. The package marking some mills employ for retrees.

XXX. The package marking some mills employ for outsides.

SUPPLIERS LIST

BEATERS

Adirondack Machine Corp.
121 Dixon Road
Glen Falls, New York 12801

Albany Felt Co.
1373 Broadway
Menands, New York 12204
Also pulp, moulds, and deckles.

American Defibrator Inc.
7400 Metro Boulevard
Minneapolis, Michigan 55435

Black Clawson Co., Shartle Div.
605 Clark Street
Middletown, Ohio 45042

Bolton Emerson Inc.
9 Osgood Street
Lawrence, Massachusetts 01842

Brodhead & Garrett Co.
4560 East 71st Street
Cleveland, Ohio 44105

Howard Clark
R.F.D. #2
Brookston, Indiana 47923

Craftool Co.
2323 Reach Road
Williamsport, Pennsylvania 17701

Dieu Donné Press
3 Crosby Street
New York, New York 10013
Also other equipment.

Edge Wallboard Machinery Co.
930 Bondsville Road
Downington, Pennsylvania 19335

Glens Falls Plug Works
Glens Falls, New York 12801

Gockel & Co.
Box 370 103D—8000
Munich 37, West Germany

Horizon Chemicals & Equipment Inc.
Box 1428
Lake Oswego, Oregon 97034

International Corp. Ltd.
1 Craven Park
London NW10 8SX England

Johnson & Cie
48/50 Rue Albert
Paris F–75013 France

Lorentzen & Wettre (AB)
Box 49006
S–10028 Stockholm, Sweden

Messmer Ltd. (H.E.)
144 Offord Road
Islington, London N1, England

St. Mary's Kraft Div.,
Gilman Paper Co.
111 West 50th Street
New York, New York 10020

Obkircher, Dipl. Ing.
Box 6044, D–7500
Karlsruhe, West Germany

David Reina
Southdown Road 246
Huntington, New York 11743

Arthur Schade
University of Wisconsin, Art Dept.
455 North Park
Madison, Wisconsin 53706

Testing Machines Inc.
400 Bayview Avenue
Amityville, New York 11701

Testing Machines International
of Canada Ltd.
6 Ronald Drive
Montreal West, Quebec
Canada H4X 1M8

Toyo Seiki, Seisaku-Sho Ltd.
Box 114, 15–4
5 Chome Takinogawa, Kita-Ku
Tokyo, Japan

Voith-Allis
P. O. Box 2337
Appleton, Wisconsin 54911

BEATER ROLLS

Diamond Inter. Corp.
Manchester Mach. Div.
P. O. Box 509
Middletown, Ohio 45042

Robert A. Main & Sons, Inc.
555 Goffle Road
Wyckoff, New Jersey 07481

Portec Inc., Cast Prod. Div.
Kingsbury Industrial Park
P. O. Box 75
Kingsbury, Indiana 46345

BED PLATES

Bolton-Emerson, Inc.
9 Osgood Street
Lawrence, Massachusetts 01842

Hamilton Industrial Grinding, Inc.
240 North B Street
Hamilton, Ohio 45013

H.J.G. McLean Ltd.
P. O. Box 202
Brantford, Ontario
Canada N3T 5M8

Portec Inc., Cast Prod. Div.
Kingsbury Industrial Park
P. O. Box 75
Kingsbury, Indiana 46345

COTTON LINTERS & VARIOUS

Alpha Cellulose Corp.
1000 East Noir Street
P. O. Box 1305
Lumberton, North Carolina 28358

Buckeye Cellulose Corp.
P. O. Box 8407
Memphis, Tennessee 38108

E. Butterworth & Co., Inc.
1951 Lakeview Avenue
Dracut, Massachusetts 01826

Castle & Overton, Inc.
1 Rockefeller Plaza
New York, New York 10020

Cheney Pulp and Paper Co.
P. O. Box 60
Franklin, Ohio 45005
250 pound minimum order.

Craftool Co.
2323 Reach Road
Williamsport, Pennsylvania 17701

Elof Hansson, Inc.
200 Park Avenue
New York, New York 10017

Hercules Inc.
910 Market Street
Wilmington, Delaware 19899

Perry H. Koplick & Sons, Inc.
654 Madison Avenue
New York, New York 10021

Elaine Koretsky
8 Evans Road
Brookline, Massachusetts 02146
Abaca and hemp pulp; sizing.

RSM Co.
900 Gwynne Bldg.
602 Main Street
Cincinnati, Ohio 45202

Simpson Timber Co.
900 Fourth Avenue
Seattle, Washington 98164
Neutral linters from redwood.

Southern Cellulose Products, Inc.
P. O. Box 7013
Chattanooga, Tennessee 37410

Sterling International
650 California Street
San Francisco, California 94108

Twinrocker
R.F.D. #2
Brookston, Indiana 47923
Wet and dry pulp, linters, etc.

Upper U.S. Paper Mill
999 Glenway Road
Oregon, Wisconsin 53575
Dry pulp.

DYES AND CHEMICALS

Aljo Manufacturing
116 Prince Street
New York, New York 10012
Dyes.

Allied Chemical Corp.
Spec. Chem. Div.
P. O. Box 1087R
Morristown, New Jersey 07960

American Color & Chemical Corp.
P. O. Box 51
Reading, Pennsylvania 19604

Amer. Cyanimid Co.
Textile & Intermed. Chem. Dept.
Bound Brook, New Jersey 08805

Amer. Hoechst Corp.
Dyes & Pigment Div.
Route 202–206 North
Somerville, New Jersey 08876

Atlantic Chemical Corp.
10 Kingsland Road
Nutley, New Jersey 07110

BASF Wyandotte Corp.
Colors & Intermed. Gp.
100 Cherry Hill Road
Parsippany, New Jersey 07054

Chemical Developments of
  Canada Ltd.
104 Doyon Avenue
Pointe Claire, Quebec
Canada H9R 3T5

CIBA-GEIGY Corp.
Dyestuffs & Chem. Div.
P. O. Box 11422
Greensboro, North Carolina 27409

City Chemical Co.
132 West 22nd Street
New York, New York 10011
Mordants, etc.

Crompton & Knowles Corp.
Ind. Prod. Div.
7535 Lincoln Avenue
Skokie, Illinois 60076

The Crystal Tissue Co.
Middletown, Ohio 45042
Dyes and papers.

Dharma Trading Co.
P. O. Box 916
San Rafael, California 94902
Procion dyes.

E.I. Dupont de Nemours & Co.
Org. Chemistry Department
Dyes Division 1007 Market Street
Wilmington, Delaware 19898

Eastman Chem. Products., Inc.,
Sub. of Eastman Kodak
P. O. Box 431
Kingsport, Tennessee 37662

Fezandie & Sperrle
111 Eighth Avenue
New York, New York 10013
Dry pigments, dyes, gum arabic, etc.

Fibrec, Inc.
2795 16th Street
San Francisco, California 94130

GAF Corp., Chem. Prods.
140 West 51st Street
New York, New York 10020

Hampden Color & Chem. Co.
126 Memorial Drive
Springfield, Massachusetts 01101

Hercules, Inc.
P. O. Box 273
Portland, Oregon 97208
Paper chemicals, additives, etc.

C. Lever Co., Inc.
736 Dunks Ferry Road
Cornwells Heights, Penna. 19020

L & R Dyestuffs Corp.
50 White Street
New York, New York 10013

Mobay Chem. Corp., Dyestuff Div.
P. O. Box 385
Union, New Jersey 07083

Hazel Pearsons Handicrafts
4128 Temple City Boulevard
Rosemead, California 91770
Dyes.

E.I. du Pont Co., Inc.
1007 Market Street
Wilmington, Delaware 19898

Pylam Products Co., Inc.
95–10 218th Street
Queens Village, New York 11429

Sandoz Colors & Chems.
East Hanover, New Jersey 07936

FELTS

Albany Felt Co. of Canada, Ltd.
Westmount Street
Cowansville, Quebec, J2K 1S9

Albany Felt Co.
1373 Broadway
Menands, New York 12204

Appleton Mills
P. O. Box 1899
Appleton, Wisconsin 54911

Ascoe Felts, Inc.
Clinton Industrial Park
Clinton, South Carolina 29325

Ayers Ltd.
Princess Street
La Chute, Quebec, Canada J8H 3X8

Carborundum Co., Lockport Felt Div.
West Avenue
Newfane, New York 14108

Coleman Sales Co., Inc.
360 Lafayette Avenue
Hawthorne, New Jersey 07506

Draper Bros. Co., Inc.
28 Draper Lane
Canton, Massachusetts 02021

Huyck Canada Ltd.
Kenwood Place
Arnprior, Ontario, Canada K7S 3H8

Huyck Felt Division, Huyck Corp.
Washington Street
Rensselaer, New York 12144

Orr Felt Co.
750 South Main
Piqua, Ohio 45356

Knox Woolen Co.
33 Mechanic Street
Camden, Maine 04843

Papermake
433 Fairlawn East
Covington, Virginia 24426

Philadelphia Felt Co.
1215 Unity Street
Philadelphia, Pennsylvania 19124

Porritts & Spencer Inc.
Box 1411
Wilson, North Carolina 27893

George B. Tewes Co.
2619 East 8th Street
Los Angeles, California 90023

H. Waterburn & Sons
Oriskamy, New York 13424

Williams-Gray Co., Inc.
5514 North Davis Highway
Pensacola, Florida 32503

GUMS

Colony Import & Export
11 East 44th Street
New York, New York 10017

Stein Hall & Co.
3926 Clenwood Drive
Charlotte, North Carolina 28208

Tragacanth Import Co.
144 East 44th Street, 7th Fl.
New York, New York 10017

MISCELLANEOUS

Local hardware stores—
waterproof adhesives

Local retail lumberyards

Fabric stores—Terylene or
dacron marquisette

Local stationers or art supply stores

Local handicrafts supply stores

MIXERS & HYDROPULPERS (NEW)

Atlantic Chemical Corp.
10 Kingsland Road
Nutley, New Jersey 07110

Baker Perkins Inc.
1000 Hess Street
Saginaw. Michigan 48601

Cellier Corp.
P.O. Box 667
Weston, Massachusetts 02193

Day Mixing Div., Le Blond Inc.
4932 Beech Street
Cincinnati, Ohio 45212

Ronald Dreager
225 Younglove Avenue
Santa Cruz, California 95060
10 gallon hydropulper.

Mixing Equipment Co.
Unit of Gen'l Signal Corp.
169 Mt. Read Boulevard
Rochester, New York 14603
Lightnin' Mixer

New Advance Machinery Co.
208 East Central Avenue
Van Wert, Ohio 45891

Perry Prod. Corp.,
Div. of Perry Equip. Corp.
Mount Laurel Road
Hainesport, New Jersey 08036

Teledyne Readco
901 South Richland Avenue
York, Pennsylvania 17405

MOULDS AND DECKLES

E. Amies & Sons
co Green Papers
Hayle Mill
Maidstone, Kent, England

Craftool Co.
2323 Reach Road
Williamsport, Pennsylvania 1770l

Lee Scott McDonald
P. O. Box 264
Charlestown, MA

Kiyufusa Narita, Director
The Paper Museum
Ojekitaku, Tokyo, Japan

PAPERMAKING KITS

Dryad
Northgates
Leicester, LE1 4QR, England

The Mould & Deckle Papermill
221 Canterbury Road
Heathmont, Vic. 3135
Australia

Paperchase Products, Ltd.
216 Tottenham Court Road
London W1, England

Papermake
433 Fairlawn East
Covington, Virginia 24426

Elliott Ruben
145 Windsor Avenue
Rockville Centre, New York 11570

Tom Starke
919 Taft Avenue
Kaukauna, Wisconsin 54130

pH EQUIPMENT (PAPER INDICATORS)

Drew Chemical Corp.
701 Jefferson Road
Parsippany, New Jersey 07054

Pfaltz & Bauer, Inc.
375 Fairfield Avenue
Stamford, Connecticut 06902

PIGMENTS (INORGANIC)

BASF Wyandotte Corp.
Colors & Intermed. Gp.
100 Cherry Hill Road
Parsippany, New Jersey 07054

E.I. Dupont de Nemours & Co.,
Pigments Div.
1007 Market Street
Wilmington, Delaware 19898

C. Lever Co., Inc.
736 Dunks Ferry Road
Cornwell Heights, Penna. 19020

L & R Dyestuffs Corp.
50 White Street
New York, New York 10013

Mobay Chem. Corp.
Dyestuffs Div.
P.O. Box 385
Union, New Jersey 07083

Reichhold Chemicals, Inc.
525 North Broadway
White Plains, New York 10603

PIGMENTS SYNTHETIC ORGANIC

American Hoechst Corp.,
Dyes & Pigment Div.
Route 202–206 North
Somerville, New Jersey 08876

BASF Wyandotte Corp.
Col. & Intermed. Gp.
100 Cherry Hill Road
Parsippany, New Jersey 07054

Chemetron Corp.,
Pigments Div.
491 Columbia Avenue
Holland, Michigan 49423

Chemetron Corp.,
Chem. Prods. Div.
12555 West Higgins Road
Chicago, Illinois 60666

Chemical Developments of Canada Ltd.
104 Doyon Avenue
Pointe Claire Montreal
Quebec, Canada H9R 3T5

CEIBA-GEIGY Corp.
Dyestuffs & Chem. Div.
P. O. Box 11422
Greensboro, North Carolina 27409

E.I. Dupont de Nemours & Co.,
Pigments Div.
1007 Market Street
Wilmington, Delaware 19898

GAF Corp., Chem. Prod.
140 West 51st Street
New York, New York 10020

General Latex & Chem. Corp.
666 Main Street
Cambridge, Massachusetts 02138

Hercules, Inc.
910 Market Street
Wilmington, Delaware 19899

Hilton-Davis Div.
Sterling Drug Inc.
2235 Langdon Farm Road
Cincinnati, Ohio 45237

C. Lever Co., Inc.
736 Dunks Ferry Road
Cornwell Heights, Penna. 19020

L & R Dyestuffs Corp.
50 White Street
New York, New York 10013

REBUILT MACHINERY, VARIOUS

Albert's Machinery Inc.
P. O. Box 453
Trenton, New Jersey 08603

Norman Albin & Assoc.
4343 West Ohio Street
Chicago, Illinois 60624

Beloit Corp., Paper Machinery Div.
1 St. Lawrence Avenue
Beloit, Wisconsin 53511

Brill Equipment Co.
35-65 Jabez Street
Newark, New Jersey 07105

Bush Mfg. Co.
P. O. Box 108
Trussville, Alabama 35173

Frank Davis Co.
P. O. Box 231
Cambridge, Massachusetts 02139

Dominion Engineering Works, Ltd.
P. O. Box 220
Montreal, Quebec,
Canada H3C 2S5

Fulton Mfg. Co.,
Div. of Ross Paper Mach.
15 Burt Street
Fulton, New York 13069

Handle Attaching Machine Corp.
138 Mott Street
New York, New York 10013

A. Johnson & Co. (Canada) Ltd.
225 Montee de Liesse Road
Montreal, Quebec, Canada H4T 1P5

Madison Equipment Co.
2950 West Carroll
Chicago, Illinois 60612

Martco Inc.
3350 Yankee Road
Middleton, Ohio 45052

Dupont de Nemours & Co., Inc.
E.I. Dupont Building, DG4
Wilmington, Delaware 19898

Nu-Pako Machinery &
   Equipment Corp.
2 Wilson Street
Bluepoint, New York 11715

The O'Brien Machinery Co.
236 Green Street
Downington, Pennsylvania 19335

Osborne Paper Mill
Fulton, New York 13069

R. Osborne Assoc.
17 Hyder Street
Westborough, Massachusetts 01581

Penn Yan Machinery, Inc.
P. O. Box 52
Lake Winola, Pennsylvania 18625

Peterson Automated Paper
   Handling Systems Inc.
38–09 10th Street
Long Island City, New York 11101

Rice Barton Corp.
Box 1086
Worcester, Massachusetts 01613

Ross Paper Machinery, Inc.
265 Passaic Street
Newark, New Jersey 07104

Ryerson & Son, Inc., Joseph T.
16th and Rockwell Streets
Chicago, Illinois 60680

Union Assoc. Inc.,
Sub. of Amer. Tool & Mach. Co.
135 Falulah Street
Fitchburg, Massachusetts 01420

Willsea Works
371 St. Paul Street
Rochester, New York 14605

Zidell Explorations, Inc.,
Valve Div.
3121 S.W. Moody Avenue
Portland, Oregon 97201

Zonel Paper Machinery Corp.
Box 484
Lockport, New York 14094

SAFETY EQUIPMENT

Bausch & Lomb
67376 Bausch Street
Rochester, New York 14602
Eyeshields and respirators.

Diamond Power Spec. Corp.
Sub. of Babcock & Wilcox
P. O. Box 415
Lancaster, Ohio 43130

Edmont-Wilson Div.
Becton, Dickinson & Co.
3110 Walnut Street
Coshocton, Ohio 43812
Clothes, gloves, oxygen equipment.

General Scientific Equip. Co.
Limekiln Pike & Williams Avenue
Philadelphia, Pennsylvania 19150
Eyeshields, first aid kits, gloves.

Goodall Rubber Co.
Whitehead Road
Trenton, New Jersey 08604
First aid kits, gloves, etc.

Industrial Products Co.
21 Cabot Road
Langhorne, Pennsylvania 19047
Eyeshields, gloves, etc.

Martindale Elect. Co.
1375 Hird Avenue
Cleveland, Ohio 44107
Eyeshields, etc.

Mine Safety Appliances Co.
400 Penn Center Boulevard
Pittsburgh, Pennsylvania 15235
First aid kits, etc.

USED MACHINERY

American Graphic Arts Inc.
150 Broadway
Elizabeth, New Jersey 07206

Converting Machinery Int'l. Inc.,
Div. of Craftsmen Mach. Co.
1073 Main Street
Millis, Massachusetts 02054

Gibbs-Brower Sales Corp.
740 South Fulton Avenue
Mount Vernon, New York 10550

Jack Frank Used Printing Equipment
947 West Cullerton
Chicago, Illinois 60608

Harris Corp., Sheet Fed Press Div.
4510 East 71st Street
Cleveland, Ohio 44105

Paper Mill Plant Exchange, Inc.
25 Sylvan Road
South Westport, Connecticut 06880

Standard Paper Box
  Machine Co., Inc.
476 Broome Street
New York, New York 10013

VARIOUS

Andrews-Nelson-Whitehead
31–10 48th Avenue
Long Island City, New York 11101
*Handmade and
mould-made papers.*

Barcham Greene & Co., Ltd.
Hayle Mill, Maidstone,
Kent ME15 6XQ, England
Handmade and

Botanica Mills
P.O. Box 1253
Fort Worth, Texas 76101
*Flowers, vegetable, and
custom-made papers.*

Douglas & Sturgess
730 Bryant Street
San Francisco, California 94107
Moulage for cast paper moulds.

Farnsworth & Co.
2325 3rd Street No. 406
San Francisco, California 94107
*Handmade and
custom-made papers.*

Graphic Chemical & Ink Co.
P.O. Box No. 27
Villa Park, Illinois 60181
General supplies.

HMP Papers
Woodstock Valley,
Connecticut 06281
*Handmade and
custom-made papers.*

Hollinger Corp.
3810 South Four-Mile Run Drive
Arlington, Virginia 22206
Acid-free matboards.

Imago
1333 Wood Street
Oakland, California 94607
*Handmade and
custom-made papers.*

Matagiri
Mt. Tremper, New York 12457
*East Indian handmade
paper samples.*

New York Central Art Supply Co.
62 Third Avenue
New York, New York 10013
*Handmade paper, papyrus,
artists supplies.*

Minnesota Clay Co.
8001 Grand Avenue
Bloomington, Minnesota 55420
Jiffy mixers.

Paperchase Products Ltd.
215 Tottemham Court Road
London W1, England
Handmade papers.

Papeterie Saint-Giles
Saint-Joseph-de-la-Rive
Comté Charlevoix
Quebec, Canada
Handmade paper.

Process Materials Corp.
329 Veterans Boulevard
Carlstadt, New Jersey 07072
*Archival materials, methyl
cellulose, pH neutral mats.*

Talas Division of Tech.
  Library Service
105 Fifth Avenue
New York, New York 10011
Archival materials.

Twinrocker, Inc.
R.F.D. No. 2
Brookston, Indiana 47923
*Handmade and
custom-made papers.*

The Two Rivers Paper Co.
Rosebank Mill
Stubbins, Nr. Bury
Lancashire, England
Handmade paper.

WIRE CLOTH (METAL OR PLASTIC)

Appleton Wire Div.,
Albany Int'l. Corp. (M & P)
P. O. Box 1939
Appleton, Wisconsin 54911

Atlanta Wire Works Inc. (PL)
1117 Battle Creek Road
Joneboro, Georgia 30236

Belleville Wire Cloth Co. (M & P)
135 Little Street
Belleville, New Jersey 07109

Capital Wire Div., Albany Int'l Co. (P)
Hinton Avenue
Ottawa, Ontario, Canada K1Y 4J6

Estey Wire Works
137 West Central Boulevard
Palisades Park, New Jersey 07650

Filtra Fabrics, Inc.
Box 205, R.D. 3
Troy, New York 12180

Holyoke Wire Cloth Co. (M & P)
650 Race Street
Holyoke, Massachusetts 01040

Huyck Formex
Div. of Huyck Corp. (P)
P. O. Box 330
Greeneville, Tennessee 37743

Johnson Wire Weaving
Div. of JWI Ltd. (M & P)
530 De Courcelle Street
Montreal, Quebec, Canada H4C 3C3

Multi-Metal Wire Cloth, Inc. (M & P)
501 Route 303
Tappan, New York 10983

Newark Wire Cloth (P)
351 Verona Avenue
Newark, New Jersey 07104

J. J. Plank Co.
Box 617
Appleton, Wisconsin 54911

H.M. Spencer Co.
78 North Canal Street
Holyoke, Massachusetts 01040

C.E. Tyler Industrial Products
Mentor, Ohio 44060

United Costa Corp. (M & P)
301 So. Livingston Avenue
Livingston, New Jersey 07039

BIBLIOGRAPHY

Adams, Alice. "Douglass Howell." *Craft Horizons* 22: 26–29. September, 1962.

Adams, W. Claude. *History of Papermaking in the Pacific Northwest*. Portland: Binfords & Mort Pub., 1951.

Ainsworth, John H. *Paper The Fifth Wonder*. Kaukauna, Wisconsin: Thomas Printing & Pub., Ltd., 1959. (3rd rev., 1967)

Adrosko, Rita J. *Natural Dyes in the United States*. Washington: Smithsonian Institution Press, 1968.

Albigny, Paul d'. *Les Industries du département de l'Ardèche*. La Papetrie, Privas, 1875.

Alibaux, Henri. *Les Premières Papeteries Françaises*. Paris, 1926.

Albright, Thomas. "The Spirit of Santa Cruz." *Art News* 75: 52–56. January, 1976.

Allemagne, Henry-René d'. *Les Cartes à Jouer du XIVe au XXe Siècle*. 2 Vols. Paris, 1906.

Allen, A. "Creativity in Paper." *Graphis*, Vol. 29, No. 166: 148–155, 1973–74.

Allen, Dot. *John Mathew, Papermaker*. London: Hodder and Stoughton Ltd., 1948.

Alloway, Lawrence. "Michelle Stuart: A Fabric of Significations." *Artforum* 12: 64–65. January, 1974.

American Cyanimid Co. *Dyestuff Data for Paper Makers*. Bound Brook, New Jersey: American Cyanimid Co., 1952.

American Paper and Pulp Association. *The Dictionary of Paper...* 3rd ed. New York, 1965.

Andés, Louis Edgar. *The Treatment of Paper for Special Purposes*. Transl. fr. German by Charles Salter. London, 1907.

Ashe, Sydney W. *Pioneer Paper Makers of Berkshire*. Dalton, Massachusetts, 1928.

Audin, Marius. *Contribution à L'histoire de la papeterie en France*. Vol IX. Grenoble, 1943.

Bayley, Harold. *The Lost Language of Symbolism*. New York, 1951.

Beatty, William B. "Early Papermaking in Utah." *The Paper Maker*, Vol. 28, No. 1, 9–20, 1959.

Beech, W.F. *Fiber Reactive Dyes*. New York: SAF International, Inc., 1970.

Bell, Jane. "Drawing Now." *Artsmagazine* 50: 6. June, 1976.

Billinger, R.D. "Early Pennsylvania Paper Making." *Journal of Chemical Education* 26: 154–158, 162. March, 1949.

Blanchet, Augustin. *Essai sur l'Histoire du Papier....* Paris, 1900.

Blum, André. *La Route du Papier*. Grenoble, 1946.

———. "First Paper Factories of the West." *Burlington Magazine* 60: 314. June, 1932.

———. *On the Origin of Paper*. Tr. by H.M. Lydenberg. New York: R.R. Bowker Co., 1934.

Bockwitz, Hans Heinrich. *Zur Kulturgeschichte des Papiers*. Stettin, 1935.

Bolam, Francis, ed. *Stuff Preparation for Paper and Paperboard Making*. London: Pergamon Press, 1965.

Bourdon, David. "Pulp Artists Paper MOMA." *Village Voice*. New York, August 23, 1976.

Bowman, Francis F. Jr. *Paper in Wisconsin....* 1940.

Bradley, Willian Aspenwall (Dard Hunter). *The Etching of Figures*. Marlborough-on-Hudson, 1915.

Briquet, Charles-Moïse. *Les Filigranes. Dictionnaire historique des marques du papier dès leur apparition vers 1282 jusqu'en 1600*. 4 Vols. Geneva, 1907.

———. *Recherches sur les premiers papiers employés en Occident et en Orient, du Xe au XIVe siècle*. Mémoires de la Societé nationale des Antiquaires de France, Vol. 46, 1886.

Browning, Bertie Lee. "Analyses of Paper." Rev. by J.C. Williams. *Stud Conserv* 17: 90. May, 1972.

———. *Analysis of Paper*. New York: Marcel Dekker, Inc., 1977.

Buck, Sir Peter H. *Arts and Crafts of Hawaii*. (See Hiroa, Te Rangi.)

Bullock, Warren B. *Romance of Paper*. Boston: Chapman and Grimes, Inc., 1933.

Butler, Frank O. *The Story of Paper-Making....* Chicago: J.W. Butter, 1901.

Cannabis. *A Report of the Commission of Inquiry into the NonMedical Use of Drugs*. Ottawa, 1972.

Carter, Thomas and Goodrich, L.C. *The Invention of Printing in China and its Spread Westward*, 2nd ed. New York: Ronald Press, 1955.

Catlin, Dana. "The Great Cases of Detective Burns." *McClures Magazine*, March, 1911, 542–556.

Charpentier, Paul. *Le Papier*. Paris, 1890.

Ch'ien, Ts'un-hsun. *Written on Bamboo and Silk. . . .* Chicago: The University of Chicago Studies in Library Science, 1962.

Chiera, Edward. *They Wrote on Clay*. Chicago: University of Chicago Press, 1938.

Clapperton, R.H. *Paper, An Historical Account. . . .* Oxford: Shakespeare Head Press, 1934.

———. *Modern Papermaking*. Oxford: Blackwell, 1952.

———. *The Papermaking Machine*. Oxford: Pergamon Press, 1967.

Cobbett, William. *A Treatise on Cobbett's Corn*. London, 1828.

Coleman, D.C. *The British Paper Industry 1495–1960. . . .* Oxford: The Clarendon Press, 1958.

Conradi, A. *Die chinesischen Handschriften-Funde Hedins*. Stockholm, 1920.

Cottier, Elie. *Histoire d'un vieux métier*. Clermont-Ferrand, 1943.

Craig, J. "How Paper is Made for Artists." *American Artist* 37: 46–49. August, 1973.

———. "Creativity with Colored Paper." *Print 16*: 16–19. July, 1962.

Crollard, Albert. *Le papier pour les livres*. Paris, 1917.

Cullum Ridgewell. *The Man in the Twilight*. New York: Putnam's Sons, 1922.

Davidson, Paul B. "Any Old Rags to Sell." *Scientific Monthly* 69: 254–261. October, 1949.

Davis, Charles Thomas. *The Manufacture of Paper. . . .* Philadelphia: Henry Carey Baird & Co., 1886. Reprint ed., Arno Press Inc., 1972.

Dawe, Edw. A. *Paper and Its Uses. . . .* London: Technical Press, 2 Vols., 1939.

Degaast, Georges and Germaine Rigaud. *Les Supports de la Pensée . . . .* Paris, 1942.

Dolloff, Francis and Roy L. Perkinson. *How to Care for Works of Art on Paper*. Boston: Museum of Fine Arts, 1971.

Donnelly, Florence. "A Fortune from Straw." *The Paper Maker*, Vol. 19, No. 1, 11–17. 1950.

———. "Camas Paper Mill, First in Washington." *The Paper Maker*, Vol. 29, No. 2, 12–28. 1960.

———. "Oregon's Pioneer Paper Mill. . . ." *The Paper Maker*, Vol. 27, No. 1, 13–22. 1958.

———. "Pioneer Paper Mill of the West." *The Paper Maker*, Vol. 18, No. 2, 15–19. 1949.

———. "The Beautiful Mill." *The Paper Maker*, Vol. 20, No. 1, 23–32. 1951.

———. "The San Lorenzo Paper Mill." *The Paper Maker*, Vol. 22, No. 1, 11–23. 1953.

———. "Trail of Ventures." *The Paper Maker*, Vol. 25, No. 1, 17–27. 1956.

Dorst, M.C. "The Paper World of Mino." *Craft Horizons* 31: 50–51. June, 1971.

"Douglass Howell, Papermaker." *Print* 12: 29–32. July, 1958.

Dravaine, Claude, *Nouara. Chroniques d'un antique village papetier*. Paris: Ed Bossard, 1927.

Dreiss, Joseph. "Group Show." *Artsmagazine* 49: 7–8. April, 1975.

———. "William Fares." *Artsmagazine* 49: 17. May, 1975.

Dunham, Judith L. "Garner Tullis: Back to the Figure." *Artweek* 6:3. January 4, 1975.

Dunham, J. "New Forms in Paper." (Martha Jackson Gallery, New York.) *Craft Horizons* 36: 38–43+. April, 1976.

"Dupont Unmasks Old-Hat Theory of Colored Paper." *Print* 13: 45–54. July, 1959.

Elliott, Harrison. "Early Papermaking in New York State." *Paper Industry* 28: 663–664. August, 1946.

———. "Events Which Lead to the Introduction of Wood Pulp as a Papermaking Material." *Pulp and Paper Magazine of Canada* 25: 931–932. July 21, 1927.

———. "Handmade Paper." *New York Public Library Bulletin* 47: 111. February, 1943.

———. "The First Paper Mill in New York." *The Paper Maker* 22: No. 2, 25–30. 1953.

———. "The Papermaker's Hand Mould." Early American Industries Association *Chronicle* 4: 37–39. October, 1951.

———. "Romance in Paper Making." *Print*, Vol. 4, No. 2: 73–74. 1946

———. "Paper As I Make It." *American Artist* 10: 17–19. March, 1946.

Engels, Johann Adolph. *Ueber Papier und einige gegenstande der Technologie und Industrie*. Duisburg, 1808.

Fenn, Dolores. "Paper Paper Paper." *Print Collector's Newsletter* 5: 33–35. May, 1974.

Fiore, Quentin. "Paper." *Industrial Design* 5: 32–60. November, 1958.

Fox, M. R. *Vat Dyestuffs and Dyeing*.

Frame, Richard. *A Short Description of Pennsilvania. . . .* Philadelphia: William Bradford, 1692.

Fuller, Mary. "Harold Paris: Sometimes by Magic it Fits." *Current Magazine*, 24–27, 56–57. June-July, 1975.

Gasparinetti, Andrea. "Carte, Cartiere e Cartai Fabrianesi." *Il Risorgimento Grafico*, September, 1938.

Gibbon, John F. "Paris in Berkeley." *Art In America* 60: 92–94. May, 1972.

Gilbert-Rolfe, Jeremy. "Clinton Hill." *Artforum* 12: 86. December, 1973.

Gilpin, Thomas. "Memoirs of Thoma Gilpin Found Among the Papers of Thomas Gilpin Jr." *Pennsylvania Magazine of History and Biography*, Vol. 49, No. 4, 289–328. 1925.

Godenne, Willy. *Six Chansons de papetiers*. Bruxelles: W. Godenne, 1966.

"Golda Lewis: From Collage to Compage." *Craft Horizons* 30: 52–53. August, 1970.

Goodwin, Rutherford. "The William Parks Paper Mill." *Southern Pulp and Paper Journal* 4: 50, 53–54, 56, 58–59, 62, 64. October, 1941.

Grant, J. "Art and Science in Paper Manufacture." *Royal Society of Arts Journal* 89: 350–363. May 2, 1941.

Gray, Cleve. "Experiments in Three Dimensions—Michael Ponce de Leon." *Art In America* 57: 72–73. May, 1969.

Greeming, William E. *Paper Makers in Canada: a History of the Paper Makers Union in Canada*. Cornwall, Ontario: International Brotherhood of Paper Makers, 1952.

Green, J. Barcham. *One Hundred and Fifty Years of Papermaking by Hand*. Maidstone, England: J. Barcham Green, 1960.

———. *Papermaking by Hand*. Maidstone, England, 1967.

Grundy, K.W. "Role of pH in the Paper Industry." *CIBA Review* 112: 4091–2. October, 1955.

von Hagen, Victor W. *The Aztec and Maya Papermakers*. New York, 1943.

Halper, A. "Douglass Howell's Handmade Papers." *Craft Horizons* 13: 13–17. May, 1953.

Hancock, Harold B. "An American Papermaker in Europe, 1795–1801." *The Paper Maker*, Vol. 28, No. 2, 11–15. 1959.

Hanson, F.S. "Resistance of Paper to Natural Aging." *Tech Stud* 8: 167–168. January, 1940.

Harnett, Cynthia. *The Load of Unicorn*. London: Methuen, 1959.

Harrocks, T.L. "Manufacture of Handmade Drawing Paper." *Art Digest* 6: 30. August, 1932.

Harte, J.H. *Volledig Molenboek*. Gorinchem, 1849.

Hartmann, Carl. *Handbuch der Papierfabrikation*. Berlin, 1842.

Hauck, Lina. *Die Muhle im Wasgau*. Kaiserslautern: Verlag Hermann Kayser, 1948.

Heitz, Paul. *Les Filigranes avec la crosse de Bâle*. Strasbourg, 1904.

Henry, Gerrit. "The Permitting Medium: A Note on the Art of Clinton Hill." *Art International* 18: 24. Summer, 1974.

Hercules, Inc. *The Paper Maker*. Wilmington, Delaware, 1932–1970.

Hergesheimer, Joseph. *The Foolscap Rose*. New York: A.A. Knopf, 1934.

Herring, Richard. *Paper & Paper Making, Ancient and Modern*. London, 1855.

Herzberg, Wilhelm. *Papierprüfung*. Berlin, 1902.

Higham, R.R.A. *A Handbook of Papermaking*. London: Oxford University Press, 1963.

Hills, R.L. *A Brief History of Papermaking*. Manchester, England: North Western Museum of Science & Industry, n.d.

Hiroa, Te Rangi (Sir Peter H. Buck). *Arts and Crafts of Hawaii*. Honolulu: Bernice P. Bishop Museum Special Publication 45, Section 5, 1964.

"How Art Paper is Made." *Comm Art* 14: 156–163. April, 1933.

Hoffner, R.J. "History of Paper Making Materials." *Paper Trade Journal*, Vol. 52, No. 10, 48. 1911.

Howell, Douglass. "Papers for Printmaking." *Art News* 49:63. May, 1963.

———. "The Future of Paper." *Print* 14: 95–96. July, 1960.

Hoyer, Fritz. *Einführung in die Papierkunde*. Leipzig, 1941.

Hunter, Dard. *A Bibliography of Marbled Paper* (Cut from Paper Trade Journal, 1921), (7) pp. New York, 1921.

———. "Ancient Paper Making." *The Miscellany*, Vol. 2, No. 4. Kansas City, Missouri, December, 1915.

———. *A Papermaking Pilgrimage to Japan, Korea and China*. New York: Pynson Printers, 1936.

———. *A Specimen of Type. . . .* Cambridge, Massachusetts: Paper Museum Press, 1940.

———. *Before Life Began, 1883–1923*. Cleveland: The Rowfant Club, 1941.

———. *Chinese Ceremonial Paper. . . .* Chillicothe, Ohio: Mountain House Press, 1917, 1937.

———. *Chronology of American Papermaking*. Holyoke, Massachusetts: B.F. Perkins, 1948.

———. "Elbert Hubbard and 'A Message to Garcia.' " *The New Colophon* Vol. 1, Pt. 1, 27–35. 1948.

———. *Fifteenth Century Papermaking*. New York: Press of Ars Typographica, 1927.

———. *Handmade Paper and its Watermarks, a Bibliography*. New York: B. Franklin, 1967. (Reprint of 1916.)

———. *Laid and Wove*. Smithsonian Institution Annual Report, 1921. 587–593. Washington, 1922.

———. "Lost and Not Lost; Handmade Paper. . . ." *Technology Review* Vol. 42, No. 3, 109–111, 124, 126. 1940.

———. *My Life With Paper*. New York: Knopf, 1958. (Bibliography, 227–237.)

———. "Ohio's Pioneer Paper Mills." *Paper Industry* 28: 100, 102, 104. April, 1946.

———. *Old Papermaking*. Chillicothe, Ohio: Mountain House Press, 1923.

———. *Old Papermaking in China and Japan by Dard Hunter*. Litt. D., Chillicothe, Ohio: Mountain House Press, 1932.

———. *Old Watermarks of Animals.* New York, 191-.

———. "Paper for Artistic Printing. . . ." *Scientific American Supplement,* Vol. LXXXII, No. 2119. New York, 1916.

———. *Papermaking by Hand in America.* Chillicothe, Ohio: Mountain House Press, 1950.

———. *Papermaking by Hand in India.* New York: Pynson Printers, 1939.

———. *Papermaking in Indo-China.* Chillicothe, Ohio: Mountain House Press, 1947.

———. *Papermaking in Pioneer America.* Philadelphia: University of Pennsylvania Press, 1952.

———. *Papermaking in Southern Siam.* Chillicothe, Ohio: The Mountain House Press, 1936.

———. *Paper-Making in the Classroom.* Peoria, Illinois: The Manual Arts Press, c.1931.

———. *Papermaking: The History and Technique of an Ancient Craft.* New York: A.A. Knopf, 1943. Also 2d. ed.—rev. and enl., 1947.

———. *Papermaking Through Eighteen Centuries.* New York: W.E. Rudge, 1930.

———. "Peregrinations and Prospects." *The Colophon,* Pt. 7, No. 8. 1931.

———. *Primitive Papermaking. . . .* Chillicothe, Ohio: Mountain House Press, 1927.

———. *Romance of Watermarks. . . .* Cincinnati, Ohio: The Stratford Press, 1939.

———. *Samples of Handmade Paper.* Chillicothe, Ohio: 1930–194_.

———. *Some Notes on Oriental and Occidental Paper and Books.* N.p., n.d.

———. *The Dard Hunter Paper Museum.* Appleton, Wisconsin, Institute of Paper Chemistry, n.d.

———. *The Early Paper Mills of Ohio.* Reprinted from the Briquet Album, pub. by the Paper Publications Society, Hilversum, Holland 1952, n.d.

———. *The Literature of Papermaking, 1390–1800.* New York: B. Franklin, 1971. Originally pub., Chillicothe, Ohio, 1925.

———. *The Story of Early Printing. . . .* Chillicothe, Ohio: Chillicothe Newspapers Inc., 194_.

———. "Ulman Stromer, . . ." *Paper,* May 8, 1921.

———, and Pels, C. *De Papierwereld door Jan Poortenaar. . . .* 2 Vols. Naarden, Uitg., "In den Toren," 1951.

Hutchinson, Helen. "John Thomson, Canadian Pulp Pioneer." *Pulp and Paper Magazine of Canada* 61: 81–83. March, 1960.

Illig, Moritz F. *Anleitung; auf eine sichere, einfache und wohlfeile Art Papier in der Masse zu leimen.* Erbach, 1807.

Imberdis, J. *Papyrus or the Craft of Paper.* North Hills, Pennsylvania, 1961 (113 numbered copies by Henry Morris); also in 1965, titled *The Paper Makers Craft* (under 400 copies by John Mason).

Indianapolis Museum of Art. *Works on Twinrocker Handmade Paper. . . .* Indianapolis: The Museum, 1975.

Institute of Paper Chemistry. *Annotated Bibliographies.* Appleton, Wisconsin: Institute of Paper Chemistry, 1974.

Jenkins, Rhys. "Early Papermaking in England, 1495–1788," *The Library Association Record* II: Nos. 9 and 11; III. No. 5; IV, Nos. 3 and 4. 1900–1902.

Johnson, Fridolf. "Henry Morris, Papermaker and Printer." *American Artist* 31: 56–61. October, 1967.

Jones, Horatio G. "Historical Sketch of the Rittenhouse Paper Mill." *Pennsylvania Magazine of History and Biography,* Vol. 20, No. 3, 315–333. 1896.

Jugaku, Bunsho. *Paper-Making by Hand in Japan.* Tokyo: Meiji-Shobo, 1959.

Karabacek, Josef. *Das Arabische Papier. . . .* Vienna, 1887.

Kent, Norman. "Brief History of Papermaking. . . ." *American Artist* 31: 36–41+. October, 1967.

King, A. *et al.* "Use of the Archivist's Pen and Universal pH Solution for Establishing the Surface pH of Paper." *Studies in Conservation* 15: 63–64. February, 1970.

"Knit Paper." *Industrial Design* 8: 44–49. March, 1961.

Kramel, H. "Outline of a Course in Plastic Art." *Graphis* Vol. 30, No. 176: 516–529. 1974–75.

Kuh, H. "Papierwerbung: ein wettbewerb der Feldmuhle." *Gebrauchsgrafik* 40: 52–56. June, 1969.

———. "Zanders' Ten Years of Consistent Promotion for Quality Papers." *Graphis* No. 171, Vol. 30, 56–57. 1974–75.

Kunisaki, Jihei. *Kamisuki Chōhōki; A Handy Guide to Papermaking.* After Jap. ed. of 1798. Tr. Chas. E. Hamilton, Berkeley: University of California, 1948.

Labarre, E.J. *A Dictionary of Paper and Papermaking Terms with a Historical Study of Paper.* Amsterdam, 1937.

Labarre, Emil Jos. *Dictionary and encyclopaedia of paper and papermaking. . . .* 2nd ed., Amsterdam: Swets & Zeitlinger, 1952. (Also *A Supplement* by Loeber, E.J., 1967.)

Laboratory Waste Disposal Manual, Washington, D.C. Manufacturing Chemists Assoc., 1975.

Lair, Pierre-Aime. *Rapport general sur les travaux de la Societé d'Agriculture et de Commerce de Caen.* Caen, 1805.

La Lande, Joseph-Jérôme Le Francois de. *L'Art de faire le papier.* New edition. Paris, 1776.

————. *The Art of Papermaking.* Kilmurry-Sixmilebridge, County Clare, Ireland: Ashling Press, 1976. (English translation of 1761 edition.)

Latour, A. "Paper, A Historical Outline." *CIBA Review* 72: 2630–2639. February, 1949.

Laufer, Berthold. *Paper and Printing in Ancient China.* New York: Burt Franklin, 1973. (Reprint of 1931 edition.)

Lavoisne, M. *A Complete Genealogical, Historical, Chronological and Geographical Atlas.* Philadelphia: Matthew Carey, 1820.

Le Clert, Louis. *Le Papier. Recherches et notes pour servir à l'histoire du papier.* 2 Vols. Paris, 1926.

Lee, H.N. "Established Methods for Examination of Paper." *Technical Studies* 4: 3–14. July, 1935.

Lee, H.N. "Micro Mechanism of Rosin Sizing." *Technical Studies* 5: 265–266. April, 1937.

Lenz, Hans. *Mexican Indian Paper, Its History and Survival.* Mexico, D.F., 1961.

Leopold, Michael. "Ron Mallory." *Art International* 18: 63. Summer, 1974.

Leschevin, Phillippe-Xavier, et Antoine, Pierre-Joseph. *Rapports lus a L'Academie des Sciences, Arts et Belles-Lettres. . . .* Dijon, 1815.

Lewis, Golda. "Douglass Howell: Sketchbooks." *Craft Horizons* 36: 30–31. August, 1976.

Lieberman, J. Ben. *Papermaking and Manufacture of Paper Products.* Washington: Office of Industrial Resources, International Cooperation Administration, 1958.

Lippard, Lucy R. "Points of View: Stuart, De Mott, Jaquette, Graves." *From The Center.* New York: Dutton, 1976.

Lorber, Richard. "William Fares." *Artsmagazine* 50: 14. February, 1976.

Lüthi, Carl Jacob. *Das Pergament.* Berne, 1938.

Maddox, H.A. *Paper: Its History, Source, and Manufacture.* Pitman, 1945.

Mason, John. *Another Edition, including Samples Printed by the Author.* Leicester: The Twelve by Eight Press, 1967.

————. *More Papers Hand Made by John Mason.* New York: Chiswick Book Shop, 1966. (Includes 35 paper samples.)

————. *Paper Making as an Artistic Craft.* Foreword by Dard Hunter. Leicester: Twelve by Eight Press, 1963. Also London: Faber & Faber, c. 1959.

————. *Twelve Papers by John Mason.* London: n.d. (c. 1959)

Mataloni, F., and Cattaneo, Carlo. "Animated Paper." *Novum Gebrauchs* 46: 38–45. September, 1975.

Mayer, Antonio James. *Printing on Cork Cut at the City Saw Mills.* London, 1851.

Mayer, Ralph. "Permanent Paper." *American Artist* 38: 10. February, 1974.

Melville, Herman. *The Selected Writings of Herman Melville.* New York: Random House, 1952. (See "The Tartarus of the Maids.")

Miller, Harry. *On the Origin of Paper.* Tr. fr. French. New York: R.R. Bowker Co., 1934.

Minsky, R. "Papermakers Cut Up at First National Meeting." *Craft Horizons* 36: 8. February, 1976.

Monumenta Chartae Papyracae Historiam Illustrantia or Collection of Works and Documents Illustrating the History of Paper. Hilversum, Holland: Paper Publications Society, 9 Vols., 1950–1962.

Morin, Louis. "Florilege de l'Imprimerie." *Bulletin officiel des maîtres imprimeurs de France,* 27–36. Paris, 1929.

Morris, Henry. *An Exhibition of Books on Papermaking.* North Hills, Pennsylvania: The Bird & Bull Press, 1968.

————. "Confessions of an Amateur Papermaker." *The Paper Maker.* Vol. 33, No. 1. 1964.

————. *Five On Paper.* North Hills, Pennsylvania: The Bird & Bull Press, 1963.

————. "Hand Papermaking as a Hobby." *The Paper Maker,* Vol. 33. No. 2, 1964.

————. "Making Paper." *Private Libraries* 6: 32–35. April, 1965.

————. *Omnibus.* North Hills, Pennsylvania: The Bird & Bull Press, 1967.

————. *The Bird & Bull Commonplace Book.* North Hills, Pennsylvania: The Bird & Bull Press, 1971.

————. *The Paper-Maker.* North Hills, Pennsylvania: Bird & Bull Press, 1974.

Müller Haupft, Suzanne. "Zauber des Papiers, Kunstverein, Frankfurt." *Das Kunstwerk* 26: 75. May, 1973.

Munsell, Joel. *Chronology of the Origin and Progress of Paper and Paper Making.* 5th ed. Albany: J. Munsell, 1876.

Murray, John. *An Account of the Phormium Tenax; or New-Zealand Flax, Printed on Paper made from its Leaves. . . .* London, 1836.

McDevitt, Jan, and Shorr, Mimi. "Paper Part II: The Solid Scrap." *Craft Horizons* 28: 27–29. January, 1968.

McMahon, J.E. "Maker of Money Paper." *Numismatist* 72: 29–30. January, 1959.

McQuade, Walter. "It Was Only A Paper Moon." *Architectural Forum* 128: 118. January, 1968.

Narita, Kiyofusa. *A Life of Ts'ai Lun and Japanese Papermaking.* Tokyo: The Dainihon Press, 1966?

Nelson, G.A. "Review of Paper Trade Terms." *Print,* Vol. 3, No. 1: 73. 1942.

Newman, Thelma et al. *Paper as Art and Craft.* New York: Crown, 1973.

Newton, J.F. "Early American Papermakers." *Antiques* 43: 271–272. June, 1943.

Noël, André. *Les compagnons de la feuille blanche.* Paris: G.T. Rageot, 1946.

le Normand. L.-Seb. *Manuel du fabricant de papier.* Paris, 1833–34.

Norris, F.H. *Paper and Paper Making.* New York, 1952.

"On Paper." *Print* 12: 2–46. July, 1958.

"On Paper: Joel Fisher in Conversation with Simon Field." *Art and Artist* 6: 32–37. January, 1972.

Overton, John. *Paper for Book Production.* London: Cambridge University Press, 1955.

"Paper Aids and Samples." *Print* 23: 62–83. May, 1969.

"Paper Mill at Dartford." *Architectural Review* 126: 279–281. November, 1959.

Papermaking: Art and Craft. Washington, D.C.: U.S. Library of Congress, 1968.

Perreault, John. "Notes on Paper." *artscanada* 25: 12–13. April, 1968.

Phillpotts, Eden. *Storm in a Teacup.* London: Heinemann, 1919.

Pindell, Howardena. "Tales of Brave Ulysses: Alan Shields Interviewed." *Print Collector's Newsletter* 5: 137–143. January, 1975.

Print. Special annual issues on paper, published in July 1958 to 1965.

Plummer, Beverly. *Earth Presents.* New York: Atheneum, 1974.

Pourrat, Henri. *Le diable au moulin à papier.* . . . Paris: Marius Péraudeau, reprint, 1963.

Poyser, James N. *Experiments in Making Paper by Hand.* Pointe Claire, Quebec, Canada: The Author, 1975.

Proteaux, A. *Practical Guide for the Manufacture of Paper and Boards.* Philadelphia, 1866.

Reilly, Desmond. "Early Papermaking in Canada." *The Paper Maker,* Vol. 21, No. 1, 13–21. 1952.

Renker, Armin. *Das Buch vom Papier.* Leipzig, 1934.

———. *Vier und Einer. Eine Papiermacher-Geschichte.* St. Gallen: Tschudy Verlag, 1952.

Ritt, A. Hugh. *Fine Paper Making.* London: Grosvenor, Chator & Co., 1966. (Includes drawings by Meg Stevens of Abbey Paper Mills, Holywell, Flintshire.)

"Robert Rauschenberg." *Print Collector's Newsletter* 6: 137. November, 1975.

Rockefeller, George C. "Early Paper Making in Trenton. . . ." New Jersey Historical Society, *Proceedings:* 71, 24–32. January, 1953.

Rose, Barbara. "The News is Paper." *Vogue.* December, 1974.

Sax, N. Irving. *Dangerous Properties of Industrial Materials.* New York: Van Nostrand Reinhold Co., 1975.

Scarborough, Jessica. *Creating Handmade Paper.* New York: Crown, Arts and Crafts Library, 1978.

Schmidt, K.C. "Synthetic Paper." *Print* 24: 70+. July, 1970.

Sarjeant, Peter T. *Handmade Papermaking Manual.* Covington, Virginia: PaperMake, 1974.

Sarjent, Peter. *The Conservation of Research Library Collections.* Chicago: Newberry Press, 1975.

Shorter, Alfred H. *Papermaking in the British Isles, an Historical and Geographical Study.* Newton Abbot, Devon, England: David & Charles, 1971.

Sime, Ian. *Making Paper.* London: Ginn & Co., Ltd., 1972.

Sindall, R.W. *The Manufacture of Paper.* London, 1908.

Slater, J. Herbert. *How to Collect Books.* London: George Bell & Sons, 1905.

Smith, David C. *History of Papermaking in the United States.* New York: Lockwood Publishing Co., 1971.

Smith, Joseph Edward Adams. *A History of Paper.* . . . Holyoke, Massachusetts: C.W. Bryan, 1882.

Sommar. Helen G. *A Brief Guide to Sources of Fiber and Textile Information.* Washington, D.C.: Information Resources Press, n.d.

Specimens. New York: The Stevens-Nelson Paper Company, 1953.

Stedman, Ebeneezer, H. *Bluegrass Craftsman.* . . . Ed. by Frances L. Dugan and Jacqueline P. Bull. Lexington: University of Kentucky Press, 1959.

Stein, M. Aurel. *Ruins of Desert Cathay.* . . . London, 1912.

Stephenson, J. Newell. *Manufacture and Testing of Paper and Board.* New York: McGraw-Hill Book Co., Inc. 3 Vols. 1953 (Various dates.)

Stephenson, J.N., Ed. *The Manufacture of Pulp and Paper.* . . . 3rd ed. New York, 1939.

Studley, Vance. *The Art and Craft of Handmade Paper.* New York: Van Nostrand Reinhold, 1977.

Sutermeister, Edwin. *The Story of Papermaking.* Boston: S.D. Warren Co., 1954.

Swanson, John W., ed. *Internal Sizing of Paper and Paperboard.* New York: Technical Association of the Pulp and Paper Industry, TAPPI monograph series no. 33. 1971.

Sweetman, John. "Making Paper by Hand." *Appropriate Technology.* Wookey Hole Caves Ltd., Wells, Somerset. February, 1977.

Tamarind Lithography Workshop. "The Beauty and Longevity of an Original Print Depends Greatly on the Paper That Supports It." Rev. November, 1966.

Tarshis, Jerome. "Designed Merriment, Elegant Disquiet." *Art News* 74: 72–73. October, 1975.

Thomson, A.G. *The Paper Industry in Scotland*. Edinburgh, 1974.

Tindale, Thomas K., and Tindale, Harriett R. *The Handmade Papers of Japan*. Rutland, Vermont and Tokyo, 1952.

Tomlinson, Charles, ed. *Cyclopaedia of the Useful Arts & Manufactures*. London, 1857.

Trotman, E.R. *Dyeing and Chemical Technology of Textile Fibers*. London: Charles Griffin and Co., Ltd., 1970.

Vander Weele, Linda Sholund. *The Revitalization of Handmade Paper in North America*. Madison: The Author, 1975.

Vinçard, B. *L'Art du Typographe*. Paris, 1806.

Voorn, Henk. "Alexander Mitscherlich, Inventor of Sulphite Wood Pulp." *The Paper Maker* Vol. 23, No. 1, 41–44. 1954.

Voorn, Henk. "De Muze en het Papier." *Papierwereld,* Vol. 4, Nos. 7, 9, 10, and 11. 1949.

Waroff, D. "Recycled Paper." *AD* 42: 673–674. November, 1972.

"James Whatman, Father and Son." Book Review of, *Burlington Magazine* 100: 448. December, 1958.

"What Jack Eisner Does With Paper." *American Artist* 7: 20–21. April, 1943.

Watermark 74. Handbook of an Exhibition. . . London, 1974, unpaged.

Weaver, Alexander. *Paper, Wasps and Packages*. Chicago: Container Corporation of America, 1937.

Weeks, Lyman H. *A History of Paper Making in the United States, 1690–1916*. New York: Lockwood Trade Journal Co., 1916.

Weiner, Jack, and Byrne, Jerry. *Sizing of Paper*. Appleton, Wisconsin: Institute of Paper Chemistry, IPC series no. 165, 163.

Weiss, Karl T. "Das Papier in Spruch und Sprache." *Altenburger Papierer*. Vols. 9–12. 1935–1938.

Weston, Harry E. "Papermaking, An Anthology." *The Paper Maker*. Vol. 12, Nos. 1 and 2. Wilmington, Delaware, 1943.

Weygand, James Lamar. *The Weygand Tightwad Beater: Its Design and Construction*. Nappanee, Indiana: The Private Press of the Indiana Kid, 1970.

Wheelwright, William Bond. *A Glossary for the Allied Trades Printing and Paper*. Boston: The Callaway Associates, 1941.

———.."A Strange Chapter in Southern Paper Making." *The Paper Maker* 2: 2–5. 1941.

———. "The Gilpins of Wilmington; Papermakers and Inventors of the American Cylinder Machine." *The Paper Maker* 1: 6–11. 1941.

———. "Pioneers in Wood Pulp." *The Paper Maker* No. 3, 4–7. 1941.

———. "Pioneers of American Paper Making." *The Paper Maker* 1: 15–20. 1940.

———. "War and the Paper Industry in 1776." *The Paper Maker* 1: 3–6. 1942.

Wilkinson, Norman B. *Papermaking in America*. Greenville, Delaware: The Hagley Museum, 1975.

Windham, Donald, "Anne Ryan and Her Collages." *Art News* 73: 76–78. May, 1974.

Wineberg, Bruce. "Hand Papermakers Hold Conference Focused on Professional Operations." *Pulp and Paper* 82–83. April, 1976.

Young, Joseph E. "Pages and Fuses: An Extended View of Robert Rauschenberg." *Print Collector's Newsletter* 5: 25–29. May, 1974.

INDEX

Edited by Ellen Zeifer
Designed by Bob Fillie
Set in 10 point Helvetica Light